KU-481-411

This book is about the social life of monkeys, apes and humans. The central theme is the importance of social information and knowledge to a full understanding of primate social behaviour and organization. Using this perspective, the authors seek to demonstrate a continuity between human and non-human society that is often not recognized elsewhere in the literature Topics covered include an overview of the contexts of behaviour; a comparison of blind strategies and tactical decision-making; social cognition; a review of intentionalist interpretations of behaviour; kinship; language and its social implications; and the constraints of culture.

Primate Behaviour will be of interest to senior undergraduate and graduate students as well as to researchers in the fields of biological and social anthropology, primatology, psychology, behavioural science, and evolutionary biology.

WITHDRAWN

LIVERPOOL JMU LIBRARY

3 1111 00719 3434

Cambridge Studies in Biological Anthropology 12

Primate behaviour – information, social knowledge, and the evolution of culture

Cambridge Studies in Biological Anthropology

Series Editors

G.W. Lasker
Department of Anatomy, Wayne State University,
Detroit, Michigan, USA

C.G.N. Mascie-Taylor
Department of Biological Anthropology,
University of Cambridge

D.F. Roberts
Department of Human Genetics,
University of Newcastle-upon-Tyne

R.A. Foley
Department of Biological Anthropology,
University of Cambridge

Also in the series

Primate behaviour

information, social knowledge,
and the evolution of culture

DUANE QUIATT

Department of Anthropology,
University of Colorado, Denver, Colorado, USA

AND

VERNON REYNOLDS

Department of Biological Anthropology,
University of Oxford, Oxford, UK

LIVERPOOL
JOHN MOORES UNIVERSITY
AVRIL ROBARTS LRC
TITHEBARN STREET
LIVERPOOL L2 2ER
TEL. 0151 231 4022

CAMBRIDGE
UNIVERSITY PRESS

Published by the Press Syndicate of the University of Cambridge
The Pitt Building, Trumpington Street, Cambridge CB2 1RP
40 West 20th Street, New York, NY 10011-4211, USA
10 Stamford Road, Oakleigh, Melbourne 3166, Australia

© Cambridge University Press 1993

First published 1993
First paperback edition 1995

Printed in Great Britain at the University Press, Cambridge

A catalogue record for this book is available from the British Library

Library of Congress cataloguing in publication data
Quiatt, Duane D.
  Primate behaviour: information, social knowledge, and the evolution of culture / Duane Quiatt and Vernon Reynolds.
    p.   cm. – (Cambridge studies in biological anthropology)
  Includes bibliographical references and index.
  ISBN 0 521 35255 X (hardback)
  1. Primates – Behaviour.   2. Cognition in animals.   3. Social evolution.
I. Reynolds, Vernon.   II. Title.   III. Series.
QL737.P9Q49   1993
599.8′0451 – dc20   92-16239   CIP

ISBN 0 521 35255 X hardback
ISBN 0 521 49832 5 paperback

PN

Contents

Acknowledgments

Many colleagues have helped in the preparation and writing of this book. We should like to thank Nicholas Mascie-Taylor for the initial suggestion, and Alan Crowden for some helpful preliminary discussions. A number of colleagues have read the whole or parts of the book, and we are especially grateful to Phyllis Lee, James McKenna, Nicholas Allen, and Robin Dunbar. Sara Trevitt and Tracey Sanderson have been helpful editors, and Alison Litherland was a meticulous sub-editor. We thank Alice Gustafsen for her help with the artwork. Finally we are greatly indebted to Leona Rozinski and Val de Newtown for help in the preparation of the manuscript.

1 *Introduction*

Our objective in writing this book is to examine the theories and evidence that bear on the subject of primate social evolution. We draw on the efforts of a generation of field workers, and on theoretical analyses concerned with the organization and evolution of social behaviour in the order Primates. Where necessary, we also draw on laboratory studies. We write as *anthropologists* and so we are concerned with the emergence of human cultural social organization, and we attempt an elucidation of the critical path which has enabled this to happen. Our approach lays emphasis on the cognitive nature of primate social communication, the need to understand primate social interactions as founded on the transmission of information, and the role of language in enabling the formulation of new levels of complexity in social formations in humans. The entire analysis is grounded in neo-Darwinian evolutionary theory, with especial emphasis on the utility of a socioecological approach, and we use this approach in an attempt to understand the emergence of human kinship systems, arguably the backbone of human social organization.

Primates live in the real world, the world of trees and rivers, rather than in some theoretical hyperspace. In this real world that they inhabit, the extent of interference of human beings varies from great, as in the case of groups of monkeys, in, say, a Japanese monkey park or around a temple in India, to small, as in the case of field studies conducted in undisturbed rain forest. Field studies present a range of interferences: the intrusion of the observer on the behaviour of the animals, seen most clearly for instance in studies relying on 'provisionization' (feeding) for access to the animals. Besides these 'real' interferences, there are further distortions resulting from the use of human language to describe what primates do, and behind these lurks the theoretical perspective determining the observer's interpretation of what is seen, which again introduces a selectivity. We need to be aware of these biases, but not allow them to defeat the object of our enquiries. Despite them, we can draw meaningful conclusions from primate studies, distinguish between varieties of ecosystems and of social organizations in those ecosystems, and draw out a way of understanding human social evolution.

Behaviour, including social behaviour, is characterized by variability, and this variability is the material on which selection works during evolution. There are two sources of variability in primate social behaviour and social organization. (By social behaviour we mean the ways in which animals interact with one another; by social organization we refer to the network of relationships making up the structure of the group as a whole.) The first is the product of what can be called ultimate causes and is the outcome of selection at the genetic level. The second is the product of what can be called proximate constraints and is the outcome of interaction between individuals and their socioecological circumstances. Both sources of variability are inextricably intertwined in actual life, but are analytically distinct. The result of the variability is that some individuals achieve greater individual fitness than others, and if we include the effects of their behaviour on their kin, then some individuals achieve greater inclusive fitness than others.

Primate habitats are largely arboreal, but a number of primate species have succeeded well in adapting to terrestrial habitats, and it is in such circumstances that some of the more complex forms of social organization have emerged, such as the matriline-based systems of some macaque and baboon species. Pressures that have led to the evolution of such systems include the presence of predator and prey species, and the richness and distribution of food supplies. Habitat diversity has led to different solutions to the problems of finding enough food in different primate species, and we find differing ways of ordering the relations between groups and the space they occupy according not only to the distribution of preferred foods but also to seasonal factors and the presence of other groups nearby. As a result, we find primates organized into multi-male groups, one-male groups, monogamous groups, or complex fission–fusion groups, presenting us with a challenge of explanation.

Behavioural ecology, which can be broken down into sociobiology and socioecology, has largely transformed our understanding of the part played by individuals in social evolution. Let us first consider the long-term or ultimate causes underlying social behaviour. For many years there was a debate between those who favoured group selection in the explanation of social organization, and those who favoured selection at the level of the individual. (By 'group selection' we mean the idea that it is groups that compete with each other for survival during evolution; the behaviour of individuals within groups is in the interests of the group as a whole (i.e. not just in their own interests), and so the best adapted groups pass on their genes while those that are less well adapted die out. By 'individual selection' we mean that individuals compete with each other for survival during evolution, and the fittest (in the Darwinian sense: i.e. those that achieve most reproductive success) pass on most genes to the next generation, while those that are less fit pass on fewer or no genes. Groups exist and survive because it is in the survival and reproductive interests of the individuals in those groups to live together and make compromises rather than living alone.) That debate is now largely over in respect of all animal species other than human beings, the latter view having superceded the former. Explanations of social cooperation and even altruism can be derived from a basis of individual competitiveness and selfishness. The twin processes of kin and reciprocal altruism, introduced by Hamilton (1964) and Trivers (1971) respectively, have been shown to have considerable explanatory power in explaining long-lasting, kin-related groups such as those characteristic of primate species.

There remains much to be incorporated into this perspective, however. For example, the distribution of genes in populations studied by population geneticists has yet to be properly integrated with behavioural ecology, and the differential fitness of individuals in groups has yet to be integrated with patterns of gene flow across generations.

Studies of primate social behaviour at the present time focus on the nature of dyadic interactions, of competition and cooperation between individuals, of the results of these on the social organization of groups, and of how individuals in social groups come to adopt particular strategies for mating and obtaining food. In the understanding of these matters, it has become clear that primate behaviour is the outcome of a continual learning process in which individuals constantly make adjustments to other group members. An economic model, viewing behaviour in terms of costs and benefits, has proved rewarding in making it possible to tie up observed behavioural outcomes with expectations derived from sociobiological and socioecological principles. Field studies have been especially productive in yielding successful analyses, but studies in captive settings where more environmental and social factors can be brought under control are essential in deciding between rival explanations.

How can strategies of the genes be connected up with the observed actions of individuals in social settings? No simple answer has been found, or is likely to be. The problem is the pervasiveness of competition, which means that any successful strategy will eventually give rise to a rival strategy, to which it will have to respond or perish, and so on. There is thus no way in which a simple set of rules of thumb can deal with each and every situation in an intelligent species such as a primate. Instead, what seems to have evolved is a set of condition/action rules for behaviour, i.e. rules that specify action according to circumstances. Strategies, consisting of a variety of tactics, comprise nested sets, giving behaviour a great variability. In order to make sense of social behaviour and organization, we have to discover regularities, and the first priority is an effective method of description. This has to some extent been achieved by the separation into levels of interactions, relationships, and social organization (Hinde 1976). (The distinctions are as follows: interactions are the patterns of behaviour that are seen when individuals are in interaction with each other; relationships are kinds of long-term associations between individuals; and social structure is the network of relationships constituting the group as a whole.)

The question remains, however, of the nature of the communicative process within the relationships in this scheme. The traditional ethological view of communication as the transfer of information has more recently been challenged by the view that communication is essentially manipulative (Dawkins & Krebs 1978). While there is some truth in the latter view, it is not conducive to an understanding of much of the communicative process in primate society. Primates have been shown to communicate quite sophisticated information to each other about individual identity, food sources, and kinds of predators. This is perhaps because primate behaviour is cognitively organized, and also because groups are mostly based on kin (normally female, but occasionally male), between whom the benefits of communicating information outweigh the benefits of trying to manipulate one another for personal gain. Also, from the anthropological perspective of trying to explain the emergence of human social organization, an informational approach seems particularly useful.

Simulations by computer programmers of information processing in the field of artificial intelligence (AI) have led to the emergence of quite sophisticated and lifelike mechanisms, such as chunking by which Allen Newell's Soar system learns from experience, converting goal-based problem solutions into long-term memory productions (Newell 1990). Solutions to problems, once solved by the program, are short-cut next time the same problem is encountered. Primates in social groups have to solve problems all the time, and may learn by such methods. What they learn is information about the other members of the group, and their fitness is determined by the use they make of this information. The idea of an information-based patterning of social knowledge provides a basis from which to explore the evolution of the institutions of human society.

How do primates 'see', or comprehend their environments, physical and social? Like all animals, they experience the world outside through their senses, mainly vision but additionally hearing, smell and taste. For vision and hearing, monkeys, apes and humans appear to have a lot in common. In our own species, we know from psychological studies that we perceive the world around us very selectively. Of the constant bombardment of our senses by the outside world, our brains and minds select what is relevant and form coherent pictures for us. Perception, for all species, is the interpretation of sensations, making sense of what is apprehended by the sense organs, and linking this up with knowledge stored in memory.

Experimental work has shown that primates are capable of transfer of learned information across sensory modalities, e.g. from the visual to the auditory, and that this ability is greater in apes than in monkeys. Here we may perhaps find clues to the evolution of spoken from gestural language.

Social perception is well developed in primates, and they have been shown experimentally to be capable of distinguishing their own relatives from unrelated group members. This knowledge about relationships can be included in the general category of social cognition. The complexity of primates' abilities to make social distinctions is paralleled in their demonstrated mental agility in experimental tests conducted by psychologists. For example Weigl oddity tests have shown that primates have the ability to generalize and so solve novel tests on the basis of rules learned in a different setting. The fact that they have a longer latency on harder tests has been taken to indicate the occurrence of vicarious mental calculations. Experimental demonstrations of the ability of primates to rank objects in a hierarchical order indicate that in a social setting this ability will be available for the organization of social relations.

These clues about primate cognitive capacities, derived from very contrived and controlled experimental tests are of interest for present purposes because they demonstrate the cognitive underpinning of primate social life. In fact, underneath this life there lies a highly evolved intelligence capable, as we know from experimental work, of making complex predictions. The biological significance of social interactions hinges on the implications for fitness, fuller understanding of which requires that we put the cognitive social skills shown by primates into a sociobiological frame of reference.

In social life, primates appear to act intentionally with regard to the strategies they pursue, but we need to beware of concluding that their kind of intentionality is closely similar to our own. Human cognitive psychology looks at the underlying biases, prejudices and presuppositions guiding social action, but this approach is not open to primatologists who must work from an assumption, namely that the behaviour of animals is fitness-oriented. Social cognition thus requires a broad definition that does not foreclose its functions, and we suggest 'the application of intelligence to review of social information and the exploitation and management of social relationships toward attainment of proximate goals' (See Chapter 6, p. 141 for further discussion of this topic.)

Social information derives from various sources including communication with other individuals. Playback experiments in the field have shown that vervet monkeys can distinguish different meanings from slight but regular acoustic differences in alarm calls. Further, they can distinguish home-group from other-group members, and dominant from subordinate group members, by vocalization alone. Thus, besides their expressive or affective function, primate calls are laden with rather specific meanings relevant to the precise situation in which they occur.

It has further been demonstrated that not only do monkeys recognize their own relationships with others, but they can recognize the relationships between other individuals in their social group. For instance, mother–daughter pairs can be matched even from photographs of the individuals concerned. Such abilities provide clues to the ease with which primates take sides in disputes between members of their own and other matrilines.

Social knowledge can be defined as 'knowledge concerning the identity of conspecific others, the character of their behaviour as individuals, and the specific nature of their relationships with one another, including oneself' (See Chapter 6, p. 148 for further discussion.) Much knowledge useful for survival can be acquired only in the social context, though it requires careful observation over long periods of time to discover how knowledge is derived from the social setting. Social knowledge, in chimpanzees at least, includes knowledge about the self, demonstrated by mirror image experiments. This ability to objectivize the self leads to a further elaboration of the complexity of social action, with the inclusion of self-awareness and self-monitoring during social interaction. The nurturing of grievances by chimpanzees, leading to aggressive revenge at long time intervals from the causal events (de Waal 1989c) indicates a powerful sense of self and, in the context of kin alliances, allows us to link up with the human institution of the feud.

Do interactions in primate social life really need such complicated explanations? Are they not capable of being fully understood in terms of stimulus–response sequences? There is no doubt that they could be so described: any behaviour can. What is missing from such oversimplified description is the background of information review and processing that leads primates to decisions such as *when* to try for a takeover, or *whether or not* to make a sexual advance to a potential partner. Do primates 'have intentionality'? What does that mean? By the analysis of cases of what appears to be deception, Whiten & Byrne (1988a,b) have shown that primates act *as if* they were working with a high degree of intentionality. As Kummer *et al.* (1990) have pointed out, further work of a more experimental kind may shed more light on such cases.

The peculiarly difficult problems presented by the analysis of deception in primate social interactions confirm that it is impossible to prove, in natural situations, that an animal intends or intended to deceive. Our own view is that primate social action is controlled by a self-conscious intelligence which functions to contemplate alternative possible courses of action and decides on the one likely to yield the best pay-off. This view is based on our own fieldwork with primates and our trust in the interpretation of others of what they have observed. The evidence in this area is necessarily somewhat anecdotal and has been subject to considerable criticism on this count. Much of this criticism is methodological. Readers will be able to see from the examples of deception in Chapter 7 what some of the methodological problems are. One issue that emerges is how to interpret objectively sets of events by after-the-fact examination of the relationships between those events.

What, if anything, can be done to bring greater precision to this issue is a currently debated issue. One aid that can at the least render behavioural observations subject to re-analysis is the use of visual recording media. This can enable re-running of a sequence that might otherwise be a mere anecdote. Experimental manipulation of relevant variables in tightly controlled settings is probably the way forward, however. Studies of the contexts in which ever more subtle tactics of deception are learned is a further way towards clarification of the processes involved. The demonstration of deception is highly situation-dependent, another complicating factor. Regularities must be sought, and these appear to include the non-transmission of information, the inhibition of action, and the hiding of give-away clues.

The significance of deception is that it involves, in primates, the manipulation of social information, or what we have called 'knowledge in the social domain'. Not all such knowledge is manipulated, indeed for the most part it is shared, so that it comes to form the body of social knowledge the animals in a group share with each other, and can communicate with each other about. The approach to the understanding of social life solely through the perspective of selfish individualism is a partial one; it omits emphasis on reciprocity, shared knowledge and the mechanisms by which relationships are maintained. It also misses the system aspect of society – the flow of information through the group that constitutes the dynamic of its socioecological adaptation.

Three *repositories* of socially communicable information can be distinguished: 1. individual memory, 2. the group's working knowledge, and 3. culturally stored knowledge – the extra-somatic archives characteristic of but not exclusive to human society. The latter for instance include the discarded sticks used for termiting by chimpanzees, which pass on technological information, or the anvil and stones used for nut-crushing by chimpanzees, which remain *in situ* for the use of others.

The second repository – the group's working knowledge – is less tangible than the third but no less real. It arises from activities performed by the group as a whole. In humans, we have the process of oral tradition of stories and songs, found all over the world. In non-human primates examples include the cooperation of all group members to locate good food sources, to deter predators by mobbing them, to escape from predators by sharing information about their species and whereabouts, or to solve problems. The latter is especially clearly demonstrated in experiments in captive situations where, for instance, chimpanzees cooperate to hold a long, heavy branch to get over an obstacle between themselves and freedom or food.

The 'language' training, or training in symbolic communication of chimpanzees and other apes demonstrates one characteristic of social knowledge, namely the setting up of a rather narrowly defined domain of knowledge, a world in which certain things or hand movements, and they alone, constitute the information base. In this closely confined world, knowledge of significant features and relationships is acquired in the course of strongly affective interaction between an ape and his or her trainer, or, as in the case of Sue Savage-Rumbaugh's chimpanzee subjects, Sherman and Austin (Savage-Rumbaugh 1986), who have to communicate symbolically to solve problems, between ape and ape. Similar 'small worlds' have been simulated in computer representations of robotic activity, with a small number of variables set to interact with each other in semi-lifelike ways. Real life groups are more open, and the passage to humanity involves a progressive opening up of the world of communicable information.

The institutionalization of relationships involves conventionalizing of knowledge in the social domain. Language is not essential for this but presumably it is essential for the evolution of *human* society. With language, the world becomes a construction, to which knowledge of the language is the key. Without language, roles can nevertheless become differentiated in conventionalized ways, as for instance in the case of lid-lifters and non-lifters among Cayo Santiago rhesus monkeys. A surprising feature of such cases is the constancy with which roles are maintained over long time periods. The playing of roles by humans who are constrained by social and economic forces shows similar constancy. In the human case, language affords conventional names for concepts and relationships and thus enables the ramification of roles in social life. Advantages in evolution accrued to those who could implement systems of categorization of information, most particularly in the extensions and re-groupings of kin seen in kinship systems, and in the construction and implementation of marriage rules and systems of affinity.

Kinship is the most elementary of all human social structures, and was probably the first area of social life to be subject to conventionalization and categorization. In macaques, matrilines exist as functional elements in the social structure, having a primary function in relation to status. Young animals are able to acquire a status consonant with that of their matriline by virtue of support from matriline members in disputes with members of other matrilines. The matrilines themselves are in competition with each other, and if, for any reason, a matriline is weakened, its members may be attacked by those of another matriline and status reversal may then take place.

The analysis of primate and human lineage structures is often achieved by means of kinship diagrams. The conventions used in these give clues to similarities and differences. An important similarity is the fact of biological (genetic) relatedness. All members of a primate matriline are so related. Human lineages are more complex, but they too are based on biological relatedness. We assume monkeys do not 'know' they are related to each other in the sense that we know that, but their relationships show different degrees of closeness or distance, and willingness to form alliances depends on the social psychology of relationships.

In human families three processes are integrated into the psychology of relationships, first the linguistic environment, second the technological environment, third the socio-legal environment. While these are prominent in human social life, equivalents of all three processes can be found in non-human primate society. In particular, these processes depend on the transmission of information and its learning by subsequent generations. For example, it is knowledge of membership of matrilines that enables monkeys to form appropriate alliances in inter-matriline conflicts.

Patrilines are unusual in primates, but do occur, for example in chimpanzees. The principles outlined above apply in the case of patrilines, except that father–son relationships replace mother–daughter ones. Patrilines and matrilines are not the same as patrilineages and matrilineages, because they involve, in primates, members of one sex only, whereas in humans both sons and daughters are members of the patrilineage or matrilineage. In primates, the 'line' is a co-residential, face to face sub-group within the whole group, and which sex is involved depends on which sex forms the permanent core of the group and which one emigrates at puberty to find a mate elsewhere. In human societies, too, it is often the case that one sex or the other leaves home at marriage, but his or her lineage ties are maintained thereafter because of the greater persistence of lineality in the context of language-based culture.

Paternal behaviour in primates was once thought to be virtually non-existent, but is now known to exist in most species. Male care of infants is widespread, and in species living in multi-male groups such as baboons is most extensive in the case of males who have formed a lasting association with a female, where it tends to be focussed exclusively on the offspring of that female (and hence very likely of the male too). This indicates the general rule: males are concerned to promote the survival of their own (genetic) offspring. In polygynous primate species living in one-male groups,[1] such as Hanuman langurs, there is fierce inter-male competition for group leadership, associated with infanticide by takeover males as a method of ensuring paternity of the group's infants. In monogamous species such as gibbons there is, by contrast, complete cooperation between males and females in territorial defence; this is associated, because of monogamy, with high paternity confidence.

The issue of paternity remains with us in modern society, as the issue of genetic techniques in legal paternity claims bears out. In most human societies, the father of a child has certain obligations towards that child's mother and the child itself, though these obligations vary from one society to another. However, because of the extensions from the primate situation made possible by language and the growth and elaborations of human cultures, all human societies have come to make a distinction not found in other primate species: between the biological aspects of relationships and the (exclusively human) jural or institutional ones. This has happened because humans have formalized their institutions in ways other primates have not.

As a result, humans have created *structures* of lineages and linked these structures to each other by *marriage*. This institution has come not only to formalize what in non-human societies are *mating* relationships, but also to link lineages to each other and maintain their status vis-a-vis each other. Marriage, indeed, emerges as the single feature that most clearly distinguishes human from non-human primate lineage structures. It has major effects on relationships. For example it brings with it the distinction between *pater* and *genitor* or what we might call 'mother's marriage partner' and 'mother's mate'. Indeed, the existence of marriage sets up a whole new tier of relationships in human society, affinal relationships, that are not based on biological interactions even if in many cases they are coincident with them.

This salient fact has been taken by many to indicate a qualitative difference between non-human and human societies, but in this book we challenge this absolute dichotomy. Our book has tried to show that primate systems are based on a flow of information between animals, and human institutions are extensions of this information flow. The factor that has led to the proliferation of roles, statuses, sanctions and so on in human institutionalized life is language, not the institutions themselves. With language, more can be done, more detailed sub-divisions can be made, formalized and logical structures of opposition can be constructed and learned by succeeding generations. The difference is of degree not of kind.

We still need however an answer to the question of why marriage evolved, and this demands an examination of what its functions are. Anthropologists such as Levi-Strauss (1969) have noted one function, namely the linking of different, and often opposed, lineages into long-term relationships based on the exchange of spouses and goods. This approaches the answer though it does not yet reach it. The conclusion generally drawn from the above is that there are advantages to group solidarity, and marriage achieves this, more especially if all lineages in a society are linked to each other in this way. Group solidarity may, for example, have advantages in inter-group competition such as warfare.

But lineages have other functions that are not necessarily geared to the interests of the group as a whole, in particular the recruitment function also stressed by some anthropologists. By having a marriage link with another lineage, individuals are assured of spouses and indeed mates. It is thus to the advantage of individuals in allied lineages to favour their lineage allies, because in course of time they are going to depend on them for reproductive support. Marriage may have evolved because of the advantages it gave to individuals in lineages so allied. Lineage alliances are also seen in incipient form in non-human primates, but only at times of group territorial defence; marriage gives humans opportunities to establish and maintain lineage links in the normal course of everyday life.

Culture exercises constraints on human individuals, and these constraints can be seen as the typically human expression of the socioecological constraints described by students of primate societies. The elaborations arising from cultural restructuring brought with them, in due course of time, the existence of *wealth* as a form of symbolic information in human society, and the great impact of wealth on patterns of mating and marriage. Indeed, wealth and the knowledge of how to obtain it are primary objectives for individuals in many human societies, and a great many cultural constraints are concerned with the pursuit of wealth.

Marriage payments, in all the societies in which they occur, reflect the wealth of the marrying families, confirm their social status, and spread knowledge of this among all the interested parties (who attend the ceremonials). The displays of wealth at such times by the families involved is the equivalent of the status competition of primate lineages. For example, matrilines in baboons or macaques are ranked relative to each other, and their relative ranks have been shown to change over time as a result of growth or decline in numbers, being eventually resolved by threats and actual fights between the matrilines. In humans the outcome of inter-lineage status competition is achieved symbolically, without threatening or fighting.

It should not, however, be assumed that there are no differences of kind between primate and human society. There is one that is of a subtle but important kind. Whereas in primates the structures such as matri- or patrilines are emergents from the behaviour patterns of individuals, the social structures of human societies differ from this in the extent of institutional control, i.e. control by the structures themselves, over the actions of individuals. This can happen because humans have reified their institutions, given them names, and vested them with powers. There is a feed-back process: human actions are constantly re-creating society, and reification of society's structures imposes constraints on action. This creates a constant interplay between the constraints on what people want to do exercised by social conventions and the exercise of choice by those people. We can see this clearly, for instance, in the choice of marriage partner: a certain partner may be preferred, but exogamy regulations may put an absolute taboo on him or her and another partner may have to be found.

The evolution of marriage appears to have been a consequence of the intelligent extension of existing categorical distinctions between lineages to create a mechanism for linking them. From early simple forms of alliance, societies have evolved the more complex matrilineal and patrilineal systems of affinity we see today, with different underlying logics according to historical circumstance. However clearly the logic of marriage systems may be articulated in the ideal world of anthropological diagrams, in the actual one borders are fuzzy and manipulation of the system is common. In this way the structures of society are able to persist over time, suggesting they must have many advantages. Among these we can envisage the advantages of information exchange and of cooperative food-getting at times of fluctuating food supply during the Palaeolithic. The evolution of language must have been an essential part of these developments.

Language evolution has yet to be satisfactorily explained. With the emergence of hominid cultures, the institutions of society became more formalized and were given names so that it was possible to talk about them and their relationships with each other. Non-human primates can handle quite complex non-verbal social information; with language this ability was greatly refined, but the contexts were the same: kinship and status. But how did spoken language itself evolve?

Hewes (1983) has posited a two-phase model for the evolution of language, an early, gestural, sememic phase of crude communication, followed by a later, phonemic one capable of infinite refinement of shades of meaning. The advantages of phonemic language are great: the same sounds can be combined in different ways to make sound units (words) with different meanings, thus greatly increasing the number of words without increasing the number of component sounds. This is how all human languages are constructed. The advantage of *Homo sapiens* over earlier forms of hominids probably stemmed from the evolution of the mental and associated capacity to phonemicize information. All evidence pertaining to this is necessarily indirect, whether from palaeontology, archaeology, language acquisition by children, or ape language learning studies.

Human children's remarkable powers of language acquisition were attributed by Chomsky (1962) to an inherited neural capacity, which he called a Language Acquisition Device. Evidence for some kind of neural facilitation of language learning seems convincing, since no other species, including apes, has been able to master the process of language production. Caution is required, however, since bonobos (pygmy chimpanzees) such as Kanzi (Savage-Rumbaugh 1990) have shown an extraordinary capacity to learn the meaning of human words and sentences, including recognition of syntax.

Both child and captivity-reared bonobo, however, live and learn in a world in which language is spoken. The analysis of how language may have evolved before such a context existed demands especially imaginative thinking. There have been few serious efforts to think this issue through, but Hurford (1989) has emphasized the selective advantages of bi-directional Saussurean signs over other kinds of sound images. In this system of information exchange, the important features are the development of a mental matching between two individuals concerning the meaning of a sound produced by either one of them. This process of negotiation of meaning is still seen in the learning of language by children, and can be seen as an outgrowth from the well documented way that young primates learn to match a set of alarm calls to a set of predators in their environment (Cheney and Seyfarth 1990).

As to the context in which language had its most pronounced advantages in evolution, this may not have been predator avoidance at all, but rather the field of social relationships and social structure. Language gives rise to the possibility of naming – of individuals, lineages, and wider groups. Given the importance of lineages in the social lives of so many primate species, it would be astonishing if kinship were not the arena in which language evolution had its greatest advantages. Such advantages accrued to individuals and lineages that were best able to bring about structural regularities in access to mating partners. This led eventually to the evolution of marriage and lineage exogamy which remain to this time the cornerstones of social structure in all human societies. The ability to manage such kinds of socially relevant information, coded phonemically, constituted a prime selective agency during human evolution.

Note

1 Terms such as 'multi-male group' and 'one-male group' refer to tendencies by particular species in particular habitats to form the type of group referred to. They should not be regarded as absolutes. A multi-male group consists of a number of adult males and a number of adult females and their young all living together. It is somewhat mis-named since the core of such groups is the females who live together permanently, whereas males leave the group at puberty to

join another one where they find mates. A one-male group by contrast is a collection of females herded or kept together by a single adult male who (with particular exceptions) keeps other adult males away; in species with this kind of social organization, such as Hamadryas baboons or Hanuman langurs in parts of their range, there exist also all-male or 'bachelor' groups from which competitors for leadership of the one-male groups arise.

LIVERPOOL JOHN MOORES UNIVERSITY
LEARNING SERVICES

2   *The contexts of behaviour*

1. Introduction

In a laboratory in England, three rhesus monkey females have been housed for the past two years with a single adult male. Just a month ago the oldest female gave birth to an infant. The youngest of the three females, not quite three and approaching menarche, evinces a particular interest in this infant, and its mother has begun to allow her to interact with it. She is grooming it now in a corner of the cage, but the infant, restless and growing hungry, voices a protest as she turns it over and presses it face down on the floor. Immediately the mother comes across to retrieve her offspring; the young female retracts her lips and releases it. The mother sits next to her, infant at the nipple, and the young female begins to groom the mother.

On an island off Puerto Rico, a group of some 50 rhesus monkeys ranges without restriction and in daily interaction with several other groups, all descended from an ancestor population transported from India several monkey generations ago. Female AG, a member of this group, has an infant daughter less than a week old. AG is seven years old, and this is her second offspring – the first, born two summers back, did not survive infancy, and it appears that this one may not survive either. She was kidnapped yesterday by four-year-old female LU, and AG has not yet succeeded in recovering her. The problem may be that LU, who in two weeks will give birth to her own first offspring, belongs to the group's most dominant matriline; since the kidnapping she has been travelling in close company with her aunt, her four-year-old male cousin, and an unrelated adult male who is a good friend of her mother. AG, a timid individual to begin with, has been threatened away by one or another of these individuals whenever she has screwed up her courage to approach LU. LU for her part is gentle and solicitous with AG's infant, which is now almost constantly on the nipple but evidently not getting milk.

In a drive-through zoo in Texas, a group of some 70 yellow baboons occupies a ten-acre enclosure. The stream of cars provides a constant source of entertainment for its younger members, males especially. Twice a day a pick-up truck comes in, turns off the road, and two

workmen drop the tailgate to unload market leftover fruits and vegetables. After they drive out again the group organizes itself in something
like concentric circles around the pile of foodstuffs, a pattern most
pronounced at the outset, when three or four large males exercise
exclusive feeding rights. Once they have eaten their fill a wave of less
dominant animals, male and female, moves in, then, with a transition less
neatly discernible, a second mixed group – until as the pile is reduced and
scattered a more fluid pattern of movement and activity is re-established.
In marked contrast to the behaviour of the baboons is that of three
Japanese macaques, males in their prime who swagger about like a trio of
furry brigands, red-faced and invincible, roving where they will and
feeding when they please, without apparent concern for the surrounding
strangers.

In the forest above Arashiyama, on the outskirts of Kyoto, Japanese
macaques collect at a feeding station where their natural diet is supplemented with durram wheat, corn, and as an occasional treat root
vegetables. In winter especially, with acorns hard to find and spring
growth yet to appear, monkeys spend long hours in this hillside clearing,
where they huddle together for warmth in groups numbering up to ten or
twelve or, on sunny days, bask against a wall or on the roof of the small
building that serves both as a centre for caretaking activities and as a
shelter for tourists come to see the monkeys. At more or less regular
intervals a middle-aged woman goes out to a shed, fills a basket with grain
and, as she has done for years, proceeds along an arbitrary but habitual
route, strewing grain about the clearing as if she were sowing a field,
followed by a pushing, vociferous crowd of monkeys which quickly
diminishes, noise level slackening, as individuals drop off at preferred
feeding spots. Here they stuff cheek pouches with food to be chewed at
leisure once the short harvest is over. Having cropped a first little stand,
some monkeys forage elsewhere, but the pickings are lean and soon
exhausted. Given the nature of this food and the manner in which it is
distributed, time spent in aggressive food competition would seem to be
time wasted. During provisioning, little time is spent in social interaction.
Internal processing of the tough, hard-coated grain, pushing it out of
pouches to chew bit by bit, continues for half an hour or so afterward, and
this is a relatively quiet period characterized by individual activity – self-
grooming or social grooming within family and between friends, rest, or,
for the young, lone exploration and object manipulation (stone handling
is a current fad). More active social play and a more active social life in
general resumes as cheek pouches are emptied, and in due course the
alpha and beta males take up again their long-standing competition for

the company of a particular high-ranking female – who these days appears to be less interested in either of them than in a third and younger male.

On a hillside in northern India, along a road that winds up from the village below, a group of langurs takes a mid-day rest, grooming and snoozing, though some few continue to feed desultorily in a grove of scattered scrubby trees. One of the females gave birth early this morning, before light, and her infant is making the rounds, being inspected, sniffed and cursorily groomed by other females in the group. When it twists impatiently or gives a whimper of distress a new female reaches for it. Once the mother intercepts it in passage and puts it on the nipple, where it sucks contentedly, but after a while she gives it up again. A bus pulls over to the side of the road to let the engine cool, and some passengers come out for a stretch. One of them throws a few peanuts toward the nearest monkeys. They approach the bus, and other passengers look to see if they too might have treats to offer. Some of the monkeys that have been feeding on leaves descend to take advantage of these easy pickings.

In Tanzania, a female chimpanzee known to millions of human beings as Flo is in estrus; today, at the Gombe Stream study site where she and other members of her community collect to obtain banana handouts, she entertains herself and a series of male acquaintances sexually over a period of several hours. Her suitors, though ardent, display no aggression in competition for her favours; indeed, there appears to be no competition as they patiently wait their turn. Although this turn-taking is impressive to the human observer, it is by no means the only pattern of chimpanzee mating, nor is it perhaps the most important in terms of resultant pregnancies. Two or three days ago another, younger female was induced by a persistent male to accompany him away from this provisioning station that has come to function also as a centre for community social interaction. Alone together now, apart from the community and unobserved by curious students of chimpanzee behaviour, they may pursue their affair for several days more of private, undivided attention to one another.

In a rain forest in Thailand, a mating pair of lar gibbons – the male black, the female, as it happens, a buff colour in this color-dimorphic species – are perched on a horizontal branch, at a fork several metres out from the trunk, high in an emergent tree. For much of the morning they have been travelling and feeding, in company with two juveniles and a subadult (all males and all buff-colored like the female), but for the past half hour they have been grooming one another in turnabout fashion. Still higher in the crown, bouncing about at the very top of the canopy, the juveniles and subadult are engaged in a game of chase. After a while the younger juvenile descends to join the grooming adults, both of which immediately transfer their attention to him. Within a few moments, however, the adult male gives over his share in this activity and departs, following the other two youngsters into an adjacent tree. There he commences grooming the other juvenile, while the subadult resumes feeding. What may appear to be a division of parental caretaking responsibilities is not quite that, for these two are the black male's brothers. They followed him when he joined the female in her territory last year, replacing her former mate. Next year he and she will produce their first joint offspring; meanwhile, a human observer would be hard put to distinguish the behaviour and spatial arrangements of this little group from that of a more 'natural' gibbon family composed of a long-term monogamous pair and their joint offspring.

Social behaviour and observational settings
These few vignettes of course do not constitute anything like a representative sample of non-human primate social behaviour (we have left out prosimians and monkeys of the New World entirely). Given the range of diversity within the order Primates, and, as far as behaviour goes, within family, genus, species and community, not to mention individuals, it is difficult to imagine what such a representative sample of behaviour sequences might look like.

Neither do we suppose that these sequences are somehow paradigmatic of basic behaviours of, e.g., foraging, mating, parenting, etc., although certainly it is not by whim that we have chosen to represent just such activities. Nor, finally, are our examples drawn at random from the literature and from our own observations. Just the opposite; they have been chosen to illustrate certain biases associated with the observation and interpretation by humans of non-human primate behaviour. If that had not caught your attention you may wish to glance at them again. We will turn eventually to the subject of bias in observation, what to do about it and what perhaps ought not to be done about it. We should note in advance, however, that it is short-sighted to think of bias in observation and interpretation simply as deviation from methodological or logical objectivity that the good scientist can and must eliminate. Some biases are deeply rooted in the historical development of method and theory, some are inherent in a particular approach – in following, for instance, a single individual through a session of observation, as opposed to making quick periodic scans to record current activities of every visible member of the group – and some emerge from the habitual linking, out of necessity or convenience, of certain kinds of studies with certain species or settings. Whether or not there can be such a thing as bias-free perspective is best left to philosophers; the point for students of behaviour to keep in mind is simply that anyone who begins by thinking he can eradicate or correct for bias in his work will be likely to end up thinking he has done it.[1]

We wanted at the outset to describe actual interactive behaviour (almost any would serve) in enough detail to illustrate differences in both physical environmental and social organizational features of setting. That raises the question of how physical environment and social organization are connected, and our primary object in this chapter is to introduce this issue, central to modern behavioural ecology. But, first and foremost, we wanted to introduce scenes with individuals in action. Individual primates are intelligent, supple, quick to learn (because slow to develop) and, being long-lived, have a complex history of behavioural choices to refer to when it comes to making some fresh decision. Just how memory works is by no means clear, but it *is* clear that, for human beings and for other primates too, the social context of every encounter reflects what interacting individuals have learned from similar encounters in times past – and may involve active recollection and re-evaluation of past experience. Thus the impact of experience on behaviour may include experience re-worked in ways to which observers are not privy and is one of the factors which, in the case of human behaviour, complicates and may confound prediction of behaviour. How far other primates resemble us in this is difficult to say, but behaviour of individual primates in society is probably no more (but perhaps no less) predictable than is that of human beings.

What is a natural setting?

The settings of the foregoing vignettes can be characterized in brief as: 1. a small enclosure; 2. a naturally bounded island in which neither the provisioned monkeys nor many of the plant species are indigenous; 3. a temporal forest, with open glades, in which an artificial provisioning station has been situated; 4. a largely deforested habitat in which naturally available foods are supplemented by scrounging in the village and from human passers-by; 5. a woodland forest in which an artificial provisioning station has been situated; and 6. a tropical forest with no provisioning facilities, minimally disturbed by human beings.

We have avoided simple dichotomies of laboratory versus field settings, or artificial versus natural habitats. The distinction between laboratory studies on the one hand, and field observations on the other, is perhaps most meaningful from an historical standpoint. Otherwise, it involves a very simplistic categorization of habitat differences, and, if making the distinction implies as it usually does parallel differences in method (e.g., experimental versus non-experimental) and in theoretical orientation (e.g., comparative psychological versus ethological or natural historical), important cross-cutting commonalities are likely to be ignored (Easley *et al.* 1989). The variety of behaviour settings in which experimental techniques are employed cannot very well be reduced to two broad categories, and the theoretical divergences which produced European ethology and North American comparative psychology in the first half of this century have been followed by rapprochement and at least partial reunification of those separate approaches in behavioural ecology.

Perhaps, basic distinctions being evident or the urge to dichotomize too strong to ignore, we can approach the indoor versus outdoor, confined versus unconfined, laboratory versus field dichotomies from another direction and refer simply to natural versus constructed settings, but there seem to be no fewer obstacles in the way here. Chief among them is undoubtedly the question of how useful it may be, in defining what is natural, to rule out of definition the constraining influence of our own species, which is dominant and intrusive where the lives of other organisms are concerned.

'Intrusive' is the key word here, for if we consider the vignette settings in terms of the degree to which human beings intrude upon the lives of animals observed there is evidently a continuum from the caged rhesus monkeys, whose lives are affected in detail and on a day-to-day basis by human behaviour, to the gibbons, whose canopy life in this particular habitat appears to be virtually unaffected (we had better say 'not directly affected') by human beings. A continuum then – and yet it is hard to think of a reported instance of primate behaviour of which one could assert with confidence that what was observed was in *no* way influenced either by the presence of the observer or by some other aspect of the observational setting and research process. It is important to keep in mind the ubiquitous character and often subtle operation of that influence whenever one is tempted to think in dichotomous terms about the behaviour of animals in the 'wild' or in 'nature' as opposed to in 'captivity' or in an 'artificial habitat'.

Differences between life in 'captivity' and in 'nature' seem so obvious that primatologists sometimes neglect to emphasize, when reporting the behaviour of, e.g., a free-ranging but long-studied group of monkeys or apes, that their study has involved animals that have been in contact with primatologists all their lives and that individual subjects may have had one or more of those same human busy-bodies hard at their heels, off and

on, for months or even years at a stretch. In the end, although there are degrees of artifice in the settings of primate behaviour, if it is human interference that makes the difference, as by definition it seems must be the case, there are no 'natural' settings.

What is 'natural' behaviour?

Central to the study of behaviour in evolution has been the notion that there are behaviours that are characteristic of a given species, and of groups of related species, and that behaviour differs across species in ways that reflect different evolutionary paths, different contexts of selection, and different adaptations thereto. The vignettes contain instances of some basic behaviours – care-giving, feeding, mating – that differ in form both within and across species. Few would argue that these behaviours are not natural to primates, as to other sexually reproducing animals. At the same time, at least where monkeys, apes, and human beings are concerned, differences in behaviour attributable in part to learning are as natural and, in their variable expression, as distinctly representative of species or groups of species as are stereotypical components of, e.g., locomotion or communicative displays.

Connections between anatomy and behaviour, between genetic and learned elements of behaviour, and between individual (non-social) and inter-individual behaviour – all are complex and subtle, whether viewed from the standpoint of evolutionary adaptation or of actualization in one or another behaviour setting. Cercopithecine monkeys, for instance, have cheek pouches that enable them to feed rapidly, collecting and storing bite-sized portions to chew at their leisure. Colobine monkeys do not have cheek pouches but have pouched stomachs; these and other specializations of the masticatory and digestive systems enable them to break down cellulose and obtain nutrients from mature, relatively dry leaves, and also to neutralize and safely absorb some of the secondary compounds that make up part of the defence system of many plants (McKenna 1979, 1987).

The social behaviour of cercopithecine monkeys is characterized by highly protective mothering and a lasting relationship between mother and young. The former tends to discourage indiscriminate allomothering; the latter encourages matrilineal ties, producing in many species strongly cohesive matrilineal sub-groupings, at least in groups of sufficient size for this tendency to be noted. Colobine monkey mothers are much less protective of young; within a few days of birth an infant langur monkey is likely to have been held and inspected by most if not all females of its group (Jay 1962,1965). How may these differences in social behaviour be related to diet and to differences noted in the anatomy and physiology of feeding?

The diet of colobine monkeys would seem, on the face of it, less catholic than that of cercopithecines – and it probably is, at least on a day-to-day basis. However, it may be that colobine leaf-eating specializations allow these monkeys to 'shift back and forth between . . . foods [of high and low nutrition and of greater and lesser toxicity]' when confronted by changes in availability (McKenna 1987). According to McKenna, leaf-eating, coupled with the ability to eat diverse foodstuffs and to go without water for long periods of time, enables colobine monkeys to weather seasonal droughts with relatively little increase in direct competition for food. In the case of cercopithecine species, relying more consistently throughout the year on foods in patchy distribution, selective pressures associated with seasonal scarcities appear to have favoured the development of close cooperation among related females competing for food with other females (see also Wrangham 1980). This cercopithecine system of 'genealogically based social hierarchies of females supporting each other in within- and between-group feeding competition' tends to preclude the kind of safe within-group access to infants by non-relatives that characterizes langur society (McKenna 1979, 1981 and, for the passage quoted, 1987).

Cercopithecine mothers, then, are both more restrictive and more selective than are colobine mothers when it comes to permitting group associates to approach and handle infant offspring. The difference, if McKenna is correct, is conditional on dietary adaptation in evolution and is deeply rooted in divergent patterns of individual foraging, with characteristic modifications of physiology, anatomy, and social behaviour. As might be expected, if protective mothering in cercopithecine monkeys is closely associated with hierarchically organized food competiton (within and among genealogical sub-groupings), a female of low rank who relinquishes her offspring puts it at greater risk than if she were of high rank. Risk may be offset by genealogical ties (or, rather, by the accustomed close spatial association and interactional familiarity that chromosomal connections normally yield), but it is always there.

The case of AJ, in the second vignette, illustrates what can happen when the wrong female, however motherly her behaviour or presumed intentions, gets hold of an infant and will not give it up. Such an outcome would be unlikely in langur society, where female competition within groups for food is not so pronounced. In langur society, infant transfer *across* groups may be formally equivalent to that described in the vignettes for rhesus monkeys, rarer and perhaps correspondingly more risky (though this is not well established) for the infant and for the mother whose genes it carries. On the other hand, infanticide by males seems to be much less frequent in cercopithecine than in colobine groups, where it has often been reported in conjunction with an intruder's takeover of a resident male's group (Sugiyama 1965, Vogel and Loch 1984, Hrdy 1977a,b; for reviews see Hausfater and Hrdy 1984, Struhsaker and Leland 1987). Reasons no doubt are as usual several and complexly related. An important one may be the apparent greater willingness of rhesus group females to gang up on a male who initiates a vicious attack on an infant (Quiatt, unpublished observations of rhesus monkeys on Cayo Santiago 1965–1987, but see Jay 1965:239 concerning the protectiveness of langur mothers).

Given the helpless dependency of very young primates it may seem 'natural' that in all species adults, male and female but especially female, should be attracted to and under ordinary circumstances gentle in their handling of conspecific infants, similarly that mothers should attempt to protect their infants against mishap – to note this is perhaps only to remind oneself that behaviour normally serves 'natural' life functions. It is at least as natural, and at least as interesting, that species should differ in even so elemental a behaviour as maternal protection and that the differences should have important consequences. Evolutionary explanations of interspecific differences in behaviour must take into account selective factors at work; given the complex web of relationships between non-social and social behaviour, the multiple functions of even the most basic social behaviours, and the multiple participant perspectives from which those functions may be interpreted, it seems clear that even the most obvious and satisfying explanation of any given social behaviour is likely to be incomplete.

The nature of social behaviour

We suggested earlier that social behaviour in primates is 'context specific'. Should that go without saying? Who would dream, anyway, of suggesting that social behaviour is *not* context-specific? After all, social behaviour not only requires some particular combination of individuals, it also occurs in the form of sequential interaction between individuals in place and in time. It is probably better not to be sidetracked, at least to begin with, by what seems to be not quite a simple definitional issue nor yet a grand philosophical question: Where does social behaviour have its locus? (But see below, p. 47) The motor components that we can observe and record (as, similarly, in non-social behaviour such as locomotion and foraging by animals isolated from conspecifics) are of course functions of individual neural, muscular, and skeletal anatomy, affected to a great though indeterminate degree by proximate physiological responses to external stimuli. Signal elements such as sound waves, pheromones and reflected light have an existence that is external to initiating and receiving individuals, and the meaning of a given communication may be discussed in terms of some relation that is established and shaped via feedback between actor and reactor. From a communication systems standpoint, then, social behaviour is context specific in the rather limited sense that its particular form and meaning cannot well be described by attending simply to an individual emitting a signal.

Let us put social behaviour and social organizational features of setting aside for the moment and consider the behaviour of individuals from a broader standpoint. To the extent that behavioural features, like anatomical form, vary in relation to habitat, we should be able to make guesses about the character of past habitats on the basis of present behaviour with some degree of confidence, granted the always risky assumption of continuity in this relation. Similarly, that continuity assumed, we should be able to make predictions about future behaviour, and with greater confidence the nearer to the present the future that we have in mind. More to the point, where behaviour of social significance is concerned, other animals, like human beings, can predict the behaviour of conspecific associates. In general terms and from an evolutionary standpoint, behaviour does vary according to habitat. Rather than ask at the outset to what degree and how we may apply this observation to more limited settings within species-characteristic habitats, let us ask in what ways habitat can be said to govern behaviour and what it implies for the study of behaviour *per se*, let alone social interaction and social organization, to acknowledge that behaviour is governed by habitat and context.

2. Behavioural variability

For any species of organism, variability in form and behaviour can be described in terms of spatial and temporal distribution and, in the case of the latter, both across and within generations and for single individuals in their life-time. Variation in typical behaviour distributed over generations or, at a given moment, across species or populations within a species is likely to have resulted from a series of mutations with allelic replacement. Change in the behaviour of a given individual, within limits set ultimately by selection and genetic evolution, of course does not reflect alteration of that individual's genotype during its life-time. Rather, it evidences a capacity for plastic adjustment in the face of changing circumstance.

Thus we find two patterns of variability in the behaviour of individuals, the underlying causes of which are qualitatively different (genetic adaptation being directly responsible in the one case, but not in the other), so that there is no simple continuum between change measured over the short term, within individual life spans, and over the long term across generations. This situation lends itself to two rather different views of how behavioural variability correlates with that of habitat and setting: on the one hand a view of behaviour as a species-characteristic evolutionary adaptation, on the other hand a view of that same behaviour as labile organismic response to circumstance. Since the manner and degree of lability in response to external stimuli are themselves moulded to some degree by selection and adaptation, the one view of course does not negate the other. We will devote more attention in what remains of this chapter to the first view, which is basic to any discussion of habitat and of the settings and contexts of behaviour.

Behavioural variability across generations: genetic adaptation
A basic premise of modern Darwinian thinking is that the influence of heredity on structure, physiology, and behaviour at all levels (i.e., from physiological process to social interaction) is constrained by selective forces associated with particular habitats or with particular structures cross-cutting a number of habitat 'types' (Standen & Foley 1989b). Differential reproductive success and consequent increase in populations of genotypes characteristic of successful lineages constitutes genetic adaptation, manifest phenotypically in transformations of structure and process. Ideally, changes in population genotype ought to be describable on a moment-by-moment basis, in the grand scale of evolutionary time, down to the present. In practice, evolutionary moments are hard to recover and extremely difficult to order reliably in series. However, gene frequencies may change significantly over a few generations; for example, in the U.S. there has been, in conjunction with the medical progress of this century, a slight but cumulatively significant rise in the frequency of abnormal genes associated with treatable conditions such as diabetes mellitus, congenital pyloric stenosis, and congenital heart diseases that if untreated would shorten the reproductive careers of individuals affected (Stern 1973:758–9). Livingstone (1969) lists a number of cases of genetic disorders in unexpectedly high frequencies in small populations, the polymorphic distribution of which probably is best explained in terms of founder effect and gene drift or sampling error. Such changes in gene frequency over the short term, within a few generations or from one generation to the next, provide population geneticists with a simple,

powerful definition of biological evolution: evolution = change in gene frequency at locus *n*. Similarly, in this shorter time-frame to which our lives are entrained and our minds attuned, stability of gene frequencies over generations can provide a concise working definition of adaptation. There is of course a problem in moving conceptually from the well-laid-out workshop of population geneticists to the cluttered laboratories and museum storehouses of palaeontologists and natural historians. In the one, precise operational measures and elegant theoretical models of populations, in their genetic and demographic aspects, afford rigorous testing of hypotheses about the workings of selection in interaction with other mechanisms of evolutionary change. In the other, fossil bones and teeth provide bits and pieces of information about the past existence of some few species (though never, unfortunately, in any direct way about their origin or demise), but precious little information bearing on either the genetic or demographic structure of a series of populations. Indeed, since the boundaries between interbreeding populations of animals long dead cannot be established, attention is limited perforce to variable features of what is left of individuals, and to what can be inferred from those features about behaviour in life. Small wonder that palaeontologists and population geneticists sometimes wrangle with one another over the relative importance of natural selection to evolutionary theory (e.g. Gould 1980, 1989; Stebbins and Ayala 1981; Dawkins 1986). Nevertheless, if different modes of analysis are required, the same principles of evolutionary change apply, with mutation and genetic adaptation at the centre.

Behavioural variability within the lifetime of individuals
The changes which an individual undergoes in life do not entail concurrent alteration of genotype. The inherited genotype remains stable, providing an elastic developmental schedule for some of those changes (in the transformations from foetus to infant to toddler, etc.), a clock for other recurrent changes(e.g., for seasonal changes in fat deposition and for short-term reversals of certain physiological functions), a trigger or end-gate for still others (as for menopause, male balding, and other features of ageing), and spelling out various conditions for all. It is useful to distinguish, despite the ambiguity that is necessarily involved in many cases, between variability that results from 'normal' schedules of development from infancy to old age, and variability that results from situation-dependent differences in response to stress associated with environmental change (Selye 1956; Candland & Leshner 1974; review by Mason 1968). Variability of the latter kind is reflected across a wide range

LIVERPOOL
JOHN MOORES UNIVERSITY
AVRIL ROBARTS LRC
TEL. 0151 231 4022

of behavioural responses, from brief whole body exercises – moving out of the sun or out of the path of an adversary, adopting a huddling posture in a downpour, etc. – to long-term, irreversible responses closely entwined with normal processes of growth and development, such as anatomical changes associated with expansion of lung capacity in humans native to extreme high altitudes (Baker & Little 1976). Intermediate are such processes as tanning in humans, seasonal changes in the pattern and quantity of fat accumulation noted parenthetically above, and much of the behaviour attributable to learning. Fuzzy boundaries confound attempts to categorize such variability neatly, and biologists are not entirely in agreement as to which categorical terms should be applied to it. Physiologists and others concerned with internal body changes have sometimes distinguished between *acclimatization*, involving gradual, largely irreversible transformations, and *acclimation,* which is accomplished more quickly and is in theory at least reversible (Harré & Lamb 1986).

Kummer (1971), discussing differences in whole organism behaviour of primates, not acclimation and acclimatization as just defined, distinguished between *genetic* and *learned* adaptations and let it go at that. Kummer demonstrated, in a series of carefully designed experiments conjoined with naturalistic observations, that the assiduous soliciting by hamadryas male baboons of prospective female followers, conjoined with punishment when invitations are ignored, is thoroughly ingrained. Hamadryas males differ from male anubis baboons with respect to this behaviour, and, since individuals raised in isolation from adult hamadryas males still exhibit it on maturing, it appears not to be simply a product of learning. Female hamadryas on the other hand, if released into an anubis group quickly adjust to the 'freer' ways of anubis sociey, while, similarly, anubis females released in the vicinity of hamadryas males learn to be 'follower females'. Thus Kummer argued that, with respect to these counterpart behaviours, the male pattern of solicitation and punitive actions toward prospective female followers is genetically entrained while female following constitutes a learned adaptation.

Kummer's distinction between *genetic* and *learned* behavioural adaptations is a useful 'first cut' distinction to make, even though evolutionary biologists, habituated to using *adaptation* without a qualifier to denote genetic consequences of selection, may find it awkward to apply the same term, however qualified, to events specifically characterized as nongenetic (but see Harrison 1988). Moreover, to complicate matters, there is a real ambiguity in the relationship between the adaptive history of most body structures and their functional integration into the observable behavior of living organisms which is likely to survive any attempt to eliminate ambiguity absolutely by definition. We assume, for instance, that the cheek pouches of rhesus monkeys reflect genetic adaptation but would be hard put to decide to what extent that assumption applies to oral transfer of food into and out of those pouches. Nor, in the view of some (Dunbar 1988; Harrison 1988), need we confine discussion of selection and adaptation to behaviour that is genetically entrained. Perhaps the most one can do is strive for clarity of reference and context in any discussion of selection processes. That, unfortunately, will not resolve the related problem, in analysing behavioural variability, of when to apply a population model of genetic transformation as opposed to an individual model of learning or how to conjoin application of the two. Models of evolution resulting from natural selection and behavioural innovation consequent on learning are closely similar with regard to the formal dynamics of change. It is easy to confuse the two, to move in discussion back and forth from one to the other without sufficient regard for conceptual boundaries. Though infrequently remarked, such confusion almost always creeps into accounts of the evolution of behaviour and may well creep into this one – readers should be on the lookout!

3. Primate habitats

It will be useful at this point to remind readers that our main concern is for the evolution of primate social behaviour – more precisely, for the development of whatever regularities are entailed in the characteristically primate social processing of information, against which we can compare the equally complex conventions of human language and culture (Chapters 8–11). Any discussion of these must require attention at the start to basic attributes of individual morphology and behaviour and to presumed ecological constants associated with their evolution; however, there are a number of good general introductions to primate evolution and primate ecology in print today, and we will not reproduce in detail what is readily available in these (see, e.g., Jolly 1985; Richard 1985; and, for primate socioecology, contributions to Standen & Foley 1989a). Thus, though we deem it essential to outline, by way of reminder (in Fig. 2.1), the most important events of primate evolution against a background of palaeoecological change, it would defeat the purpose of this book if we were to develop that sketch in yet another survey and explanation of morphological and behavioural variability across primate species. Similarly, though the subject of this section is habitat diversity, readers who require more exhaustive description or a more detailed

Eras	Periods	Epochs	World Climate	Major evolutionary events
Cenozoic	Quaternary / 12 000 yr / 1.8	Holocene		Essentially modern fauna
		Pleistocene	Further increase in upbuilding, ice-rafting; marked climatic fluctuation; continental glaciations in the northern hemisphere	*Theropithecus* largely replaced by *Papio*; *Homo* evolves; colobines and cercopithecines in Asia but disappear from Europe (cercopithecines linger until last interglacial); New World monkeys in South America
	Tertiary / 5	Pliocene	Continued global cooling, with major glaciations; development of permanent polar icecaps in both hemispheres	*Theropithecus* baboons widespread in Africa; colobines and cercopithecines in Europe and Asia; Homininae in east and south Africa; New World monkeys in South America
	22	Miocene	Cooler and drier; grasslands spread in middle latitudes; development of major icecap on east Antarctica; major increase in mountain-building, ice-rafting	Dryopithecids, sivapithecids, and (late in the epoch) colobines in Africa, Europe and Asia; in Africa, a few fossils resembling modern prosimii; first New World Monkeys (?) in South America
	38	Oligocene	Sharp acceleration of cooling, and drying; unrestricted Circum-Antarctic current in Late Oligocene, with southerly retreat of warmer seas	Anthropoid primates in Africa (the Fayum); prosimians disappear from the fossil record in Europe, Asia (?) and North America; first appearance of primates in South America
	55	Eocene	Continuation of Paleocene trends; more rapid cooling – seasonality occurring in high latitudes	Modern orders of mammals appear; prosimian primates in Europe, Asia, and North America
		Paleocene	Mountain-building; gradual cooling, but tropical climates still widespread	First major mammalian radiation, including small mammals with primate-like features
	65 MYA = millions of years ago			
Mesozoic	Cretaceous		Uniform high temperatures, humid climate, subtropical and tropical provinces extend to high latitudes	Last dinosaurs; early placental mammals

Figure 2.1.   History of the order Primates, with major events of its evolution set in a dated context of climatic and geological changes since the Cretaceous. (Sources: Richard 1985; Jolly & Plog 1979; Conroy 1990.)

categorical comparison must look elsewhere. Our goal is to be as specific as we can about the relation between habitat diversity and, in sections to follow, variation in social organizational form and individual behaviour, but without going into detail or indulging in speculation about the ultimate origins of highly specific anatomical features and functions.

To deal briefly with ecology inevitably entails simplification, especially perhaps in categorizing and representing habitat types (Figure 2.2; see below for discussion). Although we attempt in this chapter to dissect the concept of environment, not butcher it, the dissection is necessarily crude, and readers are warned against any assumption that habitat type *per se* is critical to analysis of the relationship between organism and environment. As Standen and Foley (1989b) have pointed out:

> Modern ecological theory . . . recognizes that different habitats may often share the same characteristics, while others that are superficially similar (for example, forests or woodlands) may differ markedly in resource structure and food availability . . . Patches [units of resource variability] can vary in size, quality, density, distribution and predictability in time and space. Different resources (for example fruits and insects) may have the same properties as far as their exploitation goes, and this may override the significance of gross habitat classification. (p.127)

Patch and *resource structure* are environmental attributes which have proven useful in modelling organism/environment relationships in part because they are *not* attributes of particular habitat types. Nevertheless, it is helpful for purposes of initial representation to conceive of them as cross-cutting conventionally recognized habitat types, and it is helpful for the same purposes to begin with a broad breakdown of conventionally recognized primate habitats. The focus throughout this chapter is on how we need to conceive of primate life in the constraining circumstances of nature. The background thereby established will be helpful when it comes to asking (as we shall begin to ask in Chapter 3) how to apply accepted principles of sociogenetic logic to understanding the evolution of primate social behaviour.

Box 2.1 presents a classification of primates which represents a compromise between recent and more traditional evolutionary approaches to allocating species to higher taxonomic categories. Our classification is drawn from Martin (1990), who discusses at length the problems of classification and argues that 'a compromise solution- . . . has the merit of permitting the vital property of relative stability, thus ensuring that a readily understandable set of terms will be available for discussing (among other things) the reconstruction of phylogenetic

Box 2.1.
Classification of the order Primates
The classification is based primarily on Martin (1990), secondarily on Conroy (1990), both following Simons (1972). Fossil groups and species are excluded. Common names are given in parentheses.

Order: Primates (primates)

  Suborder: Prosimii (prosimians)
    Infraorder: Lemuriformes (lemuroids)
      Superfamily: Lemuroidea (Malagasy lemurs)
        Family: Cheirogaleidae (mouse and dwarf lemurs)
              Lemuridae (true lemurs and sportive lemurs)
              Indriidae (indri group)
              Daubentoniidae (aye-aye)
    Infraorder: Lorisiformes (lorisoids)
      Superfamily: Lorisoidea
        Family: Lorisidae (loris group)
          Subfamily: Lorisinae (loris subgroup)
              Galaginae (bushbabies)
    Infraorder: Tarsiiformes (tarsioids)
      Superfamily: Tarsioidea
        Family: Tarsiidae (tarsiers)

  Suborder: Anthropoidea (simians)
    Infraorder: Platyrrhini (New World simians)
      Superfamily: Ceboidea (New World monkeys)
        Family: Cebidae (true New World monkeys)
          Subfamily: Cebinae (capuchin and squirrel monkeys)
              Aotinae (owl and titi monkeys)
              Atelinae (spider and woolly monkeys)
              Alouattinae (howler monkeys)
              Pithecinae (sakis)
              Callimiconinae (Goeldi's monkey)
        Family: Callitrichidae (marmosets and tamarins)
    Infraorder: Catarrhini (Old World simians)
      Superfamily: Cercopithecoidea
        Family: Cercopithecidae (Old World monkeys)

Box 2.1. *Continued*

Subfamily: Cercopithecinae (guenon group)
Colobinae (leaf-monkeys)
Superfamily: Hominoidea (apes and human beings)
Family: Hylobatidae (lesser apes)
Subfamily: Hylobatinae (gibbons)
Family: Pongidae (great apes)
Subfamily: Ponginae (modern great apes)
Family: Hominidae (hominids)
Subfamily: Homininae (human beings)

relationships' (p.99; Smuts *et al.* 1987 and Conroy 1990 contain classifications which are very similar to Martin's, and Conroy presents a discussion of problems in taxonomy that is similarly wide-ranging and informative). Our concern is with social behaviour in evolution, not with reconstructing in detail phylogenetic relationships throughout the order; we are perhaps even more concerned than Martin with establishing at the outset an understandable and familiar set of terms. Those who find the approach too conservative for their taste are at least unlikely to be confused by it.

The major events of primate evolution are outlined in Figure 2.1 in broad palaeoecological context, against a background of global climatological change. As has often been noted, primate species have adapted to a wide range of environmental circumstances and yet have retained, by and large, locomotor flexibility and broad dietary competence – most frequently in conjunction with highly restrictive food preferences. This general 'readiness to adapt' is most obvious at the level of the order, and again perhaps, following an initial series of adaptive radiation of Prosimii in the Paleocene and Eocene, in the rise of the Anthropoidea, especially the Old World monkeys and great apes. But note that as we begin to qualify the initial generalization, anthropocentric bias reveals itself with a nod of approval toward what we take to be an unusual flexibility of behaviour revealed by our closer relatives. Note too that our use just now of the phrase 'readiness to adapt' conflates two features, the potential for genetic adaptation and that for behavioural flexibility, which is why we set the phrase off in inverted commas. It would be more accurate to say that behavioural flexibility has been maintained at the level of the order in three ways: first through retention of physical traits which are both primitive and general (i.e., useful in a variety of habitats and behaviour contexts); second through the development of more advanced traits which, although they reflect increasing adaptation to an arboreal habitat (as, arguably, do nails coupled with sensory pads at digit ends, increased mobility and independent control of pedal digits, changes in vision and increase in brain size relative to that of the body), still involve no loss or significant impairment of underlying general function; and, third, on the level of neural and cognitive adaptation, integrative advances which ensure that learning will provide an efficient and remarkably elastic buffer against environmental adaptation tending toward genetic special-ization. Adaptive radiation within the order has produced a variety of accommodations to locomotor substrates and foraging conditions – skeletal, as in the case of arboreal and brachiating gibbons or, conversely, in primates that are almost wholly terrestrial, such as the patas monkey or our own species; dental, as in the case especially of prosimian insectivores and or gum-eaters; and a few highly specialized adaptations of specific organs, e.g., fat-storing tails in aestivorous species of Madagascar lemurs and, still in Madagascar, the aye-aye's curiously elongated third digit, a tool for harvesting larvae of wood-burrowing grubs. (For reviews in detail of anatomical adaptations, see Clark 1959; Fleagle 1988; and Martin 1990.) A list of defining characters or trends within the order Primates (see Box 2.2) is likely to contain attributes of the first and second sort but not the third, and a conventional view of the order is one in which many species are seen as retaining a capacity for multidirectional adaptation that is buffered by a similar capacity for behavioural flexibility. This view seems to us appropriate to a discussion of social behaviour and social cognition in evolution.

The features (or evolutionary trends) listed in Box 2.2 clearly are not applicable as a set to every species of primate but provide an idealized picture of the order as a whole, one that reflects its arboreal history. Consideration of these features from an anthropological standpoint has yielded insights into differences in dentition and in the cranial and skeletal anatomy of monkeys, apes, and humans, primarily as they relate to broad differences within and between these groups in diet, predator defence, and habitual locomotion in forest and savannah habitats. To put it another way, anthropological consideration of the nature of Primates as an order is rooted finally in popular and scientific concern for understand-ing human evolution. This approach, for better or worse, tends to focus attention on a relatively few groups of species and similarly few ecological settings and relations, by and large ignoring the fact that the order is composed of a great number of diverse species – some 300 living species in over 50 genera (we give rough figures for there is no consensus, but see, e.g., Napier & Napier 1985 and Smuts *et.al.* 1987) – secondarily adapted to a wide range of habitats and resource structures.

Habitat diversity

Because many species of primates exhibit behavioural flexibility which enables them to exploit more than one habitat, or to exploit one habitat in more than one manner depending on season and circumstance, no classification of primate habitats can be keyed to primate taxonomy in a way which will satisfy all or indeed very many primatologists (see e.g. Crook & Gartlan 1966. This and similar classifications appear in reprint in Sussman 1979). Further, as we noted in the previous section, while ecologists find it useful to recognize natural communities composed of populations of plants and animals which interact in distinctive abiotic environmental contexts, classification of the natural world into such communities is unavoidably arbitrary (Standen & Foley 1989b:127; see also Richard 1985:46). When natural communities are grouped by broad similarities in vegetation structure into larger biomes, diagrammatic catalogues of those groupings may or may not reflect inherent intergradation. The one in Figure 2.2 does reflect it, while intergradation is ignored in Table 2.1's listing of world biomes presently inhabited by primates. Distribution of primate species and the general ecology of primate life in various biomes has been rather thoroughly detailed in a number of recent works (see especially Jolly 1985 and Richard 1985). We will note here the main features of habitat diversity, focusing on those that appear to be most informative concerning social behaviour. The object will not be to list every feature of the environment and discuss possible implications for behaviour but to focus on the variables most meaningful for broad comparisons across species, and organizing our outline in a way that best describes how the habitat serves as a context for interactive behaviour and group social organization.

Inorganic or abiotic features

Physical environmental constituents play an important role in determining what plant and animal species will be represented in communities characteristic of a given biome. Nutrients available to primates represented therein as well as the character of foraging and losses to predation can hardly be considered without taking into account the ways in which they are influenced by climate, topography, elevation, availability of water, hours of daylight, and so on. Climate, in conjunction with

---

Box 2.2. Evolutionary trends characteristic of the order Primates

Broadly speaking, the order Primates can be defined as a natural group of mammals distinguished by the following prevailing evolutionary trends:

1. The preservation of a generalized structure of the limbs with a primitive pentadactyly, and the retention of certain elements of the limb skeleton (such as the clavicle) which tend to be reduced or to disappear in some groups of mammals.
2. An enhancement of the free mobility of the digits, especially the thumb and big toe (which are used for grasping purposes).
3. The replacement of sharp compressed claws by flattened nails, associated with the development of highly sensitive tactile pads on the digits.
4. The progressive abbreviation of the snout or muzzle.
5. The elaboration and perfection of the visual apparatus with the development to varying degrees of binocular vision.
6. Reduction of the apparatus of smell.
7. The loss of certain elements of the primitive mammalian dentition, and the preservation of a simple cusp pattern of the molar teeth.
8. Progressive expansion and elaboration of the brain, affecting predominantly the cerebral cortex and its dependencies.
9. Progressive and increasingly efficient development of those gestational processes concerned with the nourishment of the foetus before birth.
10. Prolongation of post-natal life periods.

[N]ot all primates (even those that exist to-day) have developed all the characters which are usually cited as diagnostic of the order, or have completed their development to the same degree. And of course, still less was this the case with the earliest Primates at the beginning of the Tertiary epoch. It may also be noted that the evolutionary trends associated with the relative lack of structural and functional specialization are a natural consequence of an arboreal habitat, a mode of life which among other things demands or encourages prehensile functions of the limbs, a high degree of

Box 2.2. *Continued*

visual acuity, and the accurate control and co-ordination of muscular activity by a well-developed brain.

Both the traits and the discussion are taken verbatim from Le Gros Clark's masterful work on *The Antecendents of Man*, 3rd edn (1971). It would be difficult to improve on this characterization of the Primates as a mammalian group; subsequent modifications reflect mainly differences in emphasis on one or another feature and qualifications as to the importance of an arboreal life *per se*.

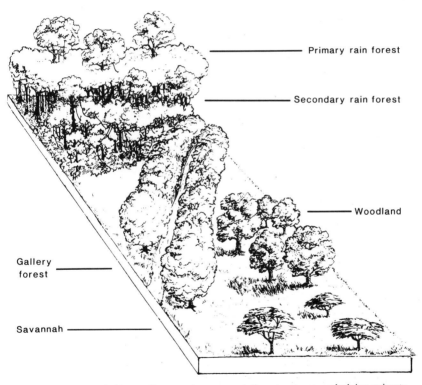

Figure 2.2. Primate habitats: diverse primary vegetation structures underlying primate communities (From Fleagle 1988, Figure 3.2).

Table 2.1. *World biomes inhabited by primates*

Tropical*	Temperate*	Subalpine and alpine	Subarctic and arctic
Rain forest	Rain forest	Elfinwood	Taiga
Seasonal or monsoon forest	Woodland, deciduous, and evergreen forest	Shrubland and meadow	Tundra
Woodland thornwood, and shrubland	Grassland		
Savanna and savanna-mosaic	Desert and semi-desert scrub		
Desert and semi-desert scrub			

*Inhabited by non-human as well as human primates.
From Richard 1985.

other factors, is of particular importance in determining what nutrients (and what predators) will be at hand, how the food supply will vary from season to season, and what structures and substrates, vegetal and inorganic, will be utilized in locating, harvesting, and processing foods.

Few would argue that climatic adaptation has not played a more direct role in the evolution of primate morphology and individual behaviour. Given the tropical origins of the order and its present distribution over a wide range of biomes one might expect to identify adaptations relating to, e.g., thermoregulation, in species whose range extends to longitudes or elevations characterized by extreme temperatures. The thick pelage of the Japanese macacque (*Macaca fuscata*) appears to be such an adaptation. However, while perhaps most features of individual morphology characteristic of this or that anthropoid species or genus (e.g., cranial morphology, dental features, and limb proportions of baboons) can be viewed as in some sense secondary or tertiary adaptations to climate, the paucity of features of distinctive function clearly attributable to selection pressures associated with high or low temperatures, humid or dry air, argues that the flexibility of acclimatization, acclimation and behaviour which we see in our own wide-ranging species is not uncommon for an anthropoid primate of Old World origin.

If it is difficult to identify features of individual morphology which represent primary adaptations to climatic stress, how much more difficult it is in the case of behaviour, especially social behaviour. In birds, thermoregulation (to continue with the same example as before) has been investigated as a factor influencing clutch size; young birds in some species depend on sibling nestmates to maintain body temperature at a viable level, radiant and convective heat loss being thus reduced through the more beneficial ratio of overall mass to air-exposed skin surface that huddling can provide. Primates of course are not birds, and for species in which singleton births are the norm this particular form of mutual insurance is in any event denied. Nevertheless, comparison is instructive. The ventral cling so characteristic of an infant primate's relation to mother serves an important heat-exchange function along with perhaps more obvious functions related to nursing and transport. For some few species of prosimians in which mothers foraging leave offspring to their own devices, multiple births and huddling of nestmates is typical. For species in which very young offspring are 'parked' in the open, selection pressure for morphological or physiological adaptation to cold stress may be greater. In the case of the smaller New World monkeys, marmosets and tamarins, characterized by twinning or triplet births, birth weight of offspring is disproportionate to adult size, and problems of heat loss may be less to begin with. It would be unwise however to conclude that thermoregulation presents no particular problems to monkeys and apes of relatively large body size. For primates of whatever size, including the most tropical of species, night-time sleeping groups are common, as is social huddling during rainstorms and at other occurrences of a drop in temperature. Where litter mates are not available and (at early critical stages of growth) conveniently located, siblings still may accommodate one another to improve heat retention on these occasions – as of course may other kin and friendly associates. The heat-seeking response of Japanese macaques in northern Japan, who, fortunately situated, retire on freezing days to hot springs, has been much publicized. Less dramatic but at least as popular, and over the long run probably more effective, is the social alternative of close huddling in groups of as many as 10 or 12 individuals.

Thermoregulation, it is clear, may depend on social behaviour in ways which are variable and complex and which differ not only across species but along individual life-spans. Study of behavioural adaptation by primates in nature to heat and cold stress, as to other features of the inorganic environment, is difficult (measures of nutrient intake and caloric expenditure are rough at best, and we are not aware of any study in which quantification of non-nutrient heat exchange has been attempted; (see Wheeler 1985); still more difficult would be to assess the role of cognition in selecting alternative strategies of behavioural adaptation (when might a Japanese monkey opt for a soak instead of a huddle, or vice versa?) Hence, perhaps, the current emphasis on social cognition in those studies of primate behaviour which deal with individual choice and the evolution of alternative strategies (Reynolds 1984). But, however difficult it may be to assess the conditioning influence of climate or other abiotic features of the environmental context of behaviour, social behaviour cannot be fully explained in terms of interactional and organizational functions.

Organization of space

Habitats are bounded horizontally by geographic features and, from the perspective of individual animals, by practical limits of day range and efficient exploitation of familiar resources, not to mention the obstructive presence on the periphery of conspecific strangers. Vertical limits of course depend on how locomotor substrates are extended in that third dimension. Internally the habitat is structured by terrain and vegetation – and, not infrequently and sometimes significantly (as in the case of urban or temple-dwelling macaques and langurs), by human habitations and constructions.

It is tempting to think of primate habitats as describable in terms of either two or three spatial dimensions, two for the more confirmed terrestrial species, e.g., Hamadryas baboons or Patas monkeys, three for those more arboreal. Making a fast and simple 'first cut' between more or less terrestrial and more or less arboreal species of primate has its uses but also sustains in an insidious way assumptions about the nature of primate life and primate evolution which are at best hard to prove. One of these assumptions is that the features listed in Box 2.2 constitute an adaptive complex which enabled early primates to cope with arboreal life in its general locomotor aspects.

An arboreal habitat in this view is a three-dimensional mazeway, to negotiate which calls for death-defying leaps or careful negotiation of tricky crossings, and which requires locomotor and path decisions in number and variety several orders beyond those that presented themselves to ground-dwelling forebears. In actuality, of course, primates have adapted to a number of *different* arboreal niches, each characterized by its own distinctive set of food species and locomotor substrates (see below). An early adaptation to one such niche may have involved insect predation in small terminal branches of the underforest, with increasing emphasis on visual location and grasping seizure of prey. In this schema, which we discuss in more detail in Chapter 5, an integration of dietary and locomotor features of behaviour is proposed (Cartmill 1974) which

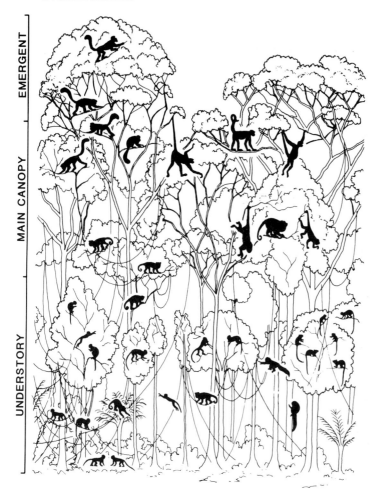

EMERGENT

MAIN CANOPY

UNDERSTORY

Figure 2.3. Locomotor substrates in a tropical rain forest (From Fleagle 1988, Figure 3.3.).

provides highly particularistic hypotheses concerning early primate evolution which can be tested against dental and anatomical features of appropriate fossil primates and palaeoecological information from relevant sites.

Locomotor substrates
The physical, three-dimensional structure of habitats consists primarily of vegetation and obtruding features of terrain – although animal organisms too, especially conspecifics (and notably mothers and play partners) can and do sometimes function as something like mobile units of physical

structure. Utilization of space depends on locomotor substrates: earth
and humus, rock outcrops, tree trunks, branches, lianas, vines and so on
are the structural members which delimit space within habitats. In the
evolution of any species there are close and obvious connections between
1. size and morphology of individuals, 2. preferred foods, and 3. substrates utilized. Comparison across species of these relations provides a
rough measure of niche separation; differences allow coexistence and
packing of related species within a single habitat (see, e.g., Hladik *et al.*
1980; for more general discussions of packing and other aspects of spatial
relationships see Dacey 1973; Hagerstrand 1973; and Quiatt *et al.* 1981;
we discuss packing of species in the next section but one, 'Interactions
across species'). How space is perceived, and what substrates are utilized
in a given sequence of behavioural action, must depend on the nature and
distribution of habitat resources – broadly defined to include, e.g., food,
water, sleeping trees, mating grounds, nest or infant-parking sites, etc. –
and, more particularly, on the immediate availability of these and on
their relevance to strategies at issue.

Conspecifics: density and distribution
For sexually reproducing animals, and especially for animals of highly
social species, a most important resource is other animals of that same
species. Interaction with conspecifics, though variable in normal frequency and character within as well as across species, is as important to
primate life as is air, water, and food. Social companions may not be
absolutely essential to survival, and it could be argued that to define life in
terms of normal interaction with conspecific others makes for a definition
rather too special and referentially too loose to mean much from a
biological standpoint. However, it is clear where our own species is
concerned that any talk of 'life' that leaves out the notion of regular,
ordinary involvement in society can't have much to do with what most of
us regard as the business of living.

   Life functions of social interaction of course are not as easy to describe
as are those of air, water, and food. We have already observed that
conspecific animals serve one another as useful physical structures. They
can be walked on, bounced off of, and ridden. On occasion they may be
eaten. Since a major concern of this book is the social uses to which
conspecific companions are put, we will content ourselves here with
noting that those companions are serviceable in other ways too many to
summarize in brief. The differential utility of particular conspecifics,
especially of genetic relatives and unrelated but familiar associates,

provides a rationale for cost/benefit decisions concerning the allocation of energetic investment in establishing and maintaining social ties (Dunbar 1984, Smuts 1985, Harcourt 1989); it provides too, from the standpoint of primatological research, a basis for systematic study of social behaviour as something more than an entertaining spectacle.

The density and distribution of conspecific animals are fairly direct functions of species-characteristic life history and demographic factors (see Standen & Foley 1989a, e.g. contributions by Harvey *et al.* and by Datta), mediated by those habitat features which have been mentioned so far and by predation pressure and interspecific competition for food resources. So perhaps are group size and occurrence of solitary individuals (Chalmers 1979), though we question whether these can be very meaningfully discussed without reference to the internal social organization of groups. We will take up discussion of that organization shortly but note here that behaviour within groups may be affected by members' individual relationships with conspecifics in other groups and with solitary animals (Chism & Rowell 1986; Cheney & Seyfarth 1990).

We have been looking at habitats as contexts for individual behaviour and social action. We are less concerned, in this chapter and from the standpoint of social behaviour and its evolution, with the particulars of habitat diversity than with asking what are the parameters of meaningful comparison. We started out by reviewing abiotic features of the environment, the general physical character of vegetation and terrain, spatial dimensions, and substrates available for foraging and other activities. Description to this point can be likened to that of a stage and settings for behaviour (cf. Hutchinson 1965). With mention of conspecifics we have come to the players themselves, and of course in evolutionary dramas these are not cast from one species alone.

Interactions across species

Social behaviour with conspecifics is affected in significant ways by interactions with non-conspecific organisms, primarily those of species definable as prey, predators, and food competitors, the last being mainly primates of other species who share the same habitat and exploit closely similar niches. While it makes sense to distinguish the three kinds of relationships and associated activities, and while for any given species of primate most other species of animals and plants can be identified in connection with one or another characteristic set of activities, there are – as is usual in nature – fuzzy boundaries. Chimpanzees and baboons are not ordinarily in close competition for food resources, but under certain circumstances they may be. When banana feeding was in its heyday at the Gombe Stream Reserve research centre, interaction between local groups of chimpanzees and baboons increased and competition became especially intense (for discussion of banana feeding see Goodall 1971, 1986a; Wrangham 1974; Reynolds 1975). Day-range and other activity patterns were altered for both species; increased competition for locally available foods, especially but not solely the introduced delicacy, can be seen as just one aspect of a general change in behaviour, though no doubt a most important aspect as far as influence on other behaviours was concerned. Agonistic behaviour directed across species was intensified (as it apparently was within species), while at the same time young chimpanzees and young baboons occasionally engaged in bouts of inter-specific play – and on a few occasions infant baboons were killed and eaten by chimpanzees (Teleki 1973; Wrangham 1977; Goodall 1986a). No case of a chimpanzee being killed and eaten by baboons has been reported, at the Gombe reserve or elsewhere, so that baboons and chimpanzees cannot be said on the basis of evidence to date to occupy all three relationship categories vis-a-vis one another. Still, it is clear that, as background to the interactional context of social behaviour within species, relations between species are no simple matter.

When we refer to 'relations between species' is it only relations with other *animal* species that we have in mind? It may seem odd to refer to behavioural relations between primates and plant species. Certainly as far as interactions go the relations between primates and plant 'prey' tend to be onesided; we perhaps could as well discuss those in a section conventionally headed 'Food resources'. Vegetal prey, after all, do not suddenly turn the tables on primate predators – there are no Venus ape- or man-traps outside Hollywood musical and space fantasy sets. On the other hand, most organisms have species-characteristic adaptations which function to defeat predators, and plants are no exception. The broader the diet, the more important it becomes to learn to feed with circumspection. Young primates must learn not only which plants but which parts of plants can be safely ingested (the same of course is true for animal prey) and in what season (the ecology of primate food relations has been treated in detail by Richard, 1985). Implications are obvious for within-species social communication and learning, as infants begin to harvest and process foods in close interaction with mothers, siblings, and other close conspecific associates.

Nor are plants harvested solely for nutritional purposes. This is painfully obvious in the case of our own species, which is rapidly exhausting the forests of the world with its incessant demand for wood products of an almost unimaginable variety, e.g., toothpicks, chopsticks, domiciles, bridge trusses, and of course newsprint. But chimpanzees and a few other species of non-human primates utilize portions of plants for tools – notably, in the case of chimpanzees, leaves for sponges and wipes, and grass stems and twigs for termiting probes. In recent years primatologists have been systematically collecting information on the medicinal use of herbs by chimpanzees and other primates (Wrangham & Nishida 1983; Huffman & Seifu 1989; Wrangham & Goodall 1989). Since non-human primates often reciprocate by disseminating (and depositing in conjunction with fertilizing faeces) the seeds of plants that they ingest, it perhaps is not entirely farfetched to think of primate/plant relationships in terms of something like behavioural interaction. For most purposes, of course, it clearly would not be very illuminating to do that, and when it comes to *social* interaction, with all that that implies, we certainly had better restrict discussion to a very few, very basic relations with certain co-feeding animal species, including other primates.

The Gombe Stream Reserve provides an especially nice instance of a chimpanzee/plant relationship deployed in social action across and then within species. In one of Jane Goodall and Hugo van Lawick's films a young juvenile male, Flint, brandishes a frond and displaces a baboon; then immediately (in film sequence at least) he repeats the display in interaction with an older chimpanzee associate who waves him off indulgently. This sort of practice is presumably useful for discovering the meaning of or, more properly said, attaching meaning to the exercise of a particular kind of behaviour (object manipulation in this instance) in a variety of contexts. Note that what is important here is not simply accumulation of experience through quantitative extension of behaviour across a range of isolated contexts, if indeed it makes sense ever to speak of experience in such terms, but what looks very much like a behavioural experiment. Whatever his intentions may be, Flint is in effect testing his frond-wielding display in two similar but far from identical social contexts which, if the spatial/temporal contiguity of those contexts is what it appears to be in the film, provide a nice opportunity for comparison of outcomes and cognitive integration of information gained.

That noted, it should be said that our concern from now on will be primarily with social behaviour *within* primate species. It is better, as a general rule, to avoid discussing the communicative aspects of interactions across species in the same terms and the same contexts of meaning as communication within species (Quiatt 1984).

4. Group social organization

Individuals of most primate species live out their lives in groups, and we therefore conceive the group to be an important influence on those lives, providing the basic social setting for individual behaviour. Size and age–sex composition of groups are functions of birth, death, and migration rates as mediated by habitat features discussed in the last section – and by typical life history and particular responses of group members to experience. For initial characterization and summary discussion of the bases of individual behaviour it is useful and not unreasonable to think of the habitat as providing a physical setting for behaviour, the group providing a social setting and a thicker interactive context. However, as we move closer to the behaving individual the awkwardness of abstracting actor and action so simply from setting becomes apparent. We will return to this point, but first let us define what we mean by the group and by social organization.

A social group of primates consists of a spatially cohesive set of individuals who recognize one another and interact on a regular and continuing basis. *Troop*, *herd*, *band*, and *community* are terms that have been applied to primate groups thus defined, where internal organization is relatively stable over the long run. Whatever term is employed is likely to require occasional qualification. For instance, it is useful when discussing the behaviour of langurs to distinguish between all-male groups and breeding groups containing both males and females, and, when discussing the behaviour of Hamadryas baboons, to distinguish between daytime foraging groups and larger sleeping groups. The virtues of a broad definition are that it does not require much qualification at these levels of inclusiveness and that it can be applied to such problematic collections as communities of chimpanzees or 'harems' of bushbabies, members of which may never all journey or gather together in one face-to-face interactive unit (Jolly 1985). The qualifications noted signal problems of interest in the study of primate behaviour, not definitional issues that have got to be resolved before there can be meaningful discussion of relations between individuals in groups. The more interesting problems of social behaviour revolve precisely around the formation and dissolution of groups, maintenance (and variable porosity) of boundaries between groups, and emergence of clusters of association and interaction within groups. When we apply an open, multi-level concept of the group to analysis of primate behaviour we perhaps are more likely to keep such problems in mind and, where it can be done, to investigate them through the construction and testing of integrative hypotheses.

Social behaviour occurs between and among individuals, as processual elements of individual behaviour which, insofar as they are learned, are learned and instantiated in interaction (cf. Savage-Rumbaugh 1986:32, on language: 'Language learning is not an individual accomplishment, but an individual process . . .'). Let us look at how patterns of individual social behaviour are developed by primatologists in the field into pictures of relationships between individuals and subsequently into models of social organization – keeping in mind, however, that the mechanics of observation and description enforce a bias of viewpoint which we may well mistake for analytic virtue simply because we can see no other way of proceeding. Visually it is practical to focus on and track only one moving object or individual at a time, and neural processing of information seems to proceed in the same way. Narrative description, necessarily serial and selective to begin with, reflects a similarly consistent focus, so that we end up reporting what actions monkey A initiated and what events or (where social behaviour is concerned) what actions initiated by other monkeys he or she reacted to. Concentrating attention on a 'focal subject' for description of his or her interactions with others over a period of record is the most popular and, for many purposes, the most productive method of sampling primate behaviour systematically in the field (Altmann 1974). From a methodological standpoint, it formalizes and justifies the habit of reporting that monkey A groomed, hit, threatened, mounted or displaced B or was groomed, hit, threatened, mounted or displaced *by* B. Furthermore, although an observer may have noted that B and C, sitting together, both got up and left when A walked in their direction, practical problems of record storage and subsequent quantitative analysis of who did what to whom often dictate that such a complex triadic event be treated as two simultaneous dyadic interactions (when indeed the presence of one or the other displaced individual is not conveniently interpreted as irrelevant to the action and left out of the account altogether, as we suspect must now and again occur).

This is not a critique of method, nor are we much concerned here with unstated philosophies of action that may underlie formulation of method. (If we were we might choose to discuss to what extent these involve western notions of property, power, and virtue imported into seemingly neutral theories of communication and behavioural exchange.) The point is simply to note that familiar practical problems of data recording and analysis (see, e.g., Altmann 1974; Slater 1973) tend to insure that however we end up thinking about social behaviour we will have started out making notes on it not as a system attribute nor as a complex statement of relations between or among individuals in time and space but as, in effect, the economic expenditure of energy by participating *individuals* toward a common behavioural product, contributed to the social system via dyadic interaction.

If group social behaviour involves the integration of individual contributions in interaction with others, as in some sense it must, understanding depends on distinguishing different patterns of input. A first step toward understanding is to establish the character of monkey A's relations with, e.g., monkey B, primarily in terms of relative frequency of different kinds of behaviour and of differences in intensity or other quality of expression depending on circumstance (Hinde 1976; Chalmers 1979). At least as important as these are *changes* in relative frequency and expressive content over time, as a consequence in part of repeated interactions and repetitive patterned sequences of behaviour. Unfortunately, sequential analysis even of very short interactive bouts is rarely pursued as part of the normal process of building up a picture of group social organization, mostly for the obvious reason that it is difficult, expensive, and enormously time-consuming (see e.g. Altmann 1965). However, it may be also that outcomes which reflect learning in response to complex sequences of particular events simply do not lend themselves to summary comparison and generalization. The immediate next step, after characterizing monkey A's interactions with monkey B, is to ask how these may differ from interactions with monkeys C, D, E, and so on, thus describing the pattern of individual relationships which obtain between monkey A and all of her or his associates (as expressed schematically in Figure 2.4). This can provide the basis for discussion of relationships involving reciprocal roles (Bernstein & Sharpe 1966; Gartlan 1968). Once individual patterns of behaviour in dyadic interaction have been described for all group members, comparison of similarities and differences within and across age/sex classes and behaviour cohorts (e.g. orphans, immigrant males, multiparous females) can yield a catalogue of roles typical of those classes and cohorts. That catalogue can then be read as a description of social organization, and a group's social organization defined thereby as the sum total of relationships and reciprocal roles that obtain within and between age/sex classes and cohorts, as expressed in behaviour typical of group members.

Such a definition may call to mind the table of organization of a business firm or military unit, with age/class substituted for civilian or military occupation, and there are similarities in that both take for granted the priority and coherence of the group and of its aims – chief among which for primate or other animal groups frequently has been

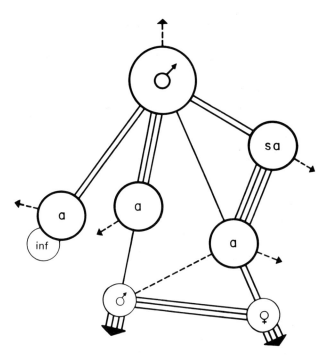

Figure 2.4.   In this sociogram of a hamadryas one-male group, the number of lines connecting the individuals represents the frequency of social interactions among them. The broken arrows indicate that the male (♂) and his adult (a) and subadult (sa) females interacted with outsiders in less than 3% of the observation minutes. In contrast, the two juveniles (bottom of graph) scored 40% in interactions with other groups. (From Kummer 1968.)

assumed to be self- perpetuation through reproduction by members. Size and age-sex composition of groups are a function of birth, death, and migration rates as mediated by habitat features discussed in the last section – and by typical life history and particular responses of group members to experience (Chalmers 1979). Until the mid–1960s primatologists and ethologists in general were not much concerned about questions of group versus individual selection. (That began to change with publication in 1962 of Wynne-Edward's *Animal Dispersion in Relation to Social Behaviour* – we will review the group- versus individual-selection issue in Chapter 3.) Field primatologists in the 1950s and 1960s, whatever other problems they may have had in mind, usually were engaged in systematically describing and comparing behaviour, that is, compiling species-characteristic 'ethograms' and comparing them across species. For these purposes the day-to-day foraging group appeared to constitute a natural unit of study, one that varied in ways that might indicate differential adaptation to habitat. Significantly, the foraging group appeared to be for many species of primates identical with the breeding group and even – until reliable information began to be obtained on emigration of individuals from their natal group (e.g. Koford 1963a,b) – with the deme itself. Description in these early studies was used to document general assertions about within-group functions of social behaviour, functions usually related to defence against predation or to reproductive competition among males or both; the general assertions were framed in terms of typical relations between age–sex and other classes, almost always either adult male/adult female or mother/offspring (see e.g., Carpenter 1942), with little attention paid to individual variation within class or to behaviour in extra- and inter-group contexts (Quiatt 1987b).

From this functionalist perspective on the analysis of group social behaviour, description of parts (or classes and typical roles) in relation to one another and in terms of some unifying whole-system purpose amounted to an outline of control features and in some sense an explanation of the system under study. Anthropologists, it has been suggested (Gilmore 1981; Richard 1985), have been unduly influenced by A.R. Radcliffe-Brown toward descriptive studies of primate behaviour emphasizing cohesion and structure of the group over individual concerns of its members. This criticism does not quite seem to take into account the very widespread and pervasive influence in the mid-century of functionalism and systems thinking on western science in general. Systems analysis and system functions explanations of complex events probably have been used more effectively in biology than in the social sciences. But, be that as it may, the light they have cast on social behaviour *in evolution* has been surprisingly dim. Perhaps this is because abstract, timeless (i.e., synchronic) analyses of structural and organizational functions lend themselves primarily to typological representations and static models of entities that are assumed implicitly to be in perpetual equilibrium after some presumed history of evolutionary adaptation. Of course, those who apply individual fitness models to explanations of ultimate causal relationships between gene and environment often make the same assumption (J. McKenna, personal communication).

When comparing groups of different organizational form it is of course necessary to simplify and generalize. In these circumstances it is difficult not to present group typologies as if they referred to static entities and to construct categorical outlines like those presented in Table 2.2, taking the spatially cohesive foraging group (=troop, band, herd, etc.) as a starting point. It is difficult to know how else one would *initiate* comparison of social organization across species, but readers are cautioned that such outlines, with their table-of-organization descriptions of primate social structure, are rough and ready, useful primarily for just that purpose of initial comparison. From this standpoint the classifications in Table 2.2 may serve as negative examples. Another approach to understanding primate social systems is to begin with the analysis of relationships between feeding, parenting, and mating behaviour.

Differences in male and female strategies of competition for resources have important implications for gene dispersal in mammalian species, in which offspring are more dependent on female parents. Female competition is likely to centre on resources crucial to offspring care, and distribution of individual females (or groups of females with overlapping reproductive interests) is thus likely to be particularly influenced by competition for food resources. Both sexes of course compete for mates; however, male–male competition for food tends to be tied rather directly to competition for mates, consequently the pattern of distribution of males is likely to be contingent on that of females, and not the other way round.

In the great majority of primate species reproductively mature males emigrate while females remain in the natal group in company with mothers, sisters, and other kin. Wrangham (1980) has suggested that where the preferred foods of a species are distributed in defensible patches it will pay females to cooperate in defending food patches from single individuals, that the natural basis for cooperation is lineal relationship, and that defence of relatively large food patches by related females, coupled with male emigration, should give rise to similarly large, generally promiscuous multi-female/multi-male groups. For species in which similarly stable groups of female relatives defend smaller patches, polygyny will prevail if groups follow the economic policy of allowing only one male to associate with them.

Wrangham's model of female-bonded (FB) group social organization provides a convincing explanation, for a great number of living species of primates, of how nutritional, reproductive, and social behavioural mechanisms are integrated in group life. It distinguishes clearly between environment and behaviour as elements influencing the character of social organization, but not as clearly between selection and learning as shaping processes. As an ecological model it differs from earlier correlational schemes in that it proceeds, not from specific correlations to general explanations, but 'from a general assertion about mammalian biology to a set of derived and more specific propositions about how male and female primates distribute themselves' (Richard 1985:379). As a purported explanation of at least some kinds of group social organization (those observable in FB species) it focuses to begin with on the behaviour of females, not males. It is important to note that 1. this focus is grounded in a (mating system) theory that is well-confirmed and broadly explanatory and 2. it takes dependent young into account in its consideration of female–female competition and, ultimately, of the organization of society. In both these respects Wrangham's model differs from the several schemes in 2.2, all of which (not incidentally) posit explicitly or by implication males as central to, and male behaviour as determinative of, group social organization. Finally, Wrangham's model has as a major virtue the fact that it distinguishes clearly between mating system features and features of more general social organization. Rowell (1988) provides a cogent warning against conflating these.

Primatologists (including anthropologists) increasingly are interested more in the social processes which underlie typological differences than in the differences *per se*, regarding the group in dynamic terms 'as but a steady-state phase of longer-term processes of troop formation, dissolution, or membership replacement' (P.C. Reynolds 1976, cited by Richards 1985). Following accumulation of information on numerous different groups of the same species and on the same groups over several generations, research interest has gradually shifted to focus on behaviour in process, on aspects of sequence, duration, intensity, etc., on relations within and between sub-groupings, and on the specific character of interactions between individuals of known genealogy and history of associations. In studies of social behaviour, the individual animal is necessarily the unit of observation and description; today the individual is likely to be the primary unit of analysis as well, with close attention paid not just to variability within and between classes of individuals but to that within the individual during his or her life-time.

5. Proximate mechanisms and ultimate design of behaviour

To what extent does this variability of behaviour displayed by an individual reflect choice? Or, to put the question more strongly, are we justified in perceiving the behaviour of nonhuman primates as we do usually that of human beings, in terms of volition and action, sometimes at least (and especially where social behaviour is concerned) intentional in the most meaningful sense of the term, involving deliberate review of circumstances and the development and enactment of some considered strategy in response? The answer to that question is, in our opinion, Yes –

but to justify that opinion we must anticipate our answer with a prior question.

What are the causes of behaviour?

David McFarland, wondering why it is that birds sit on eggs, has applied this question to four different problems: Why do birds sit on eggs rather than on other objects? Why do they sit on eggs instead of doing something else with eggs? Why do birds sit on eggs while cats and pigs do not sit on eggs? and What is it that birds accomplish by sitting on eggs? (MacFarland 1985, pp. 2–3; see also Tinbergen 1963 for the classic statement of these questions.) The first question seeks to define the stimuli which elicit egg-sitting, the second deals with motivation, the third requires discussion of hereditary predisposition, and the fourth directs attention to function. The first two questions 'require answers in terms of immediate cause or mechanism', the last two require discussion of evolutionary design of egg-sitting, i.e., the logic of inheritance and of selection and adaptation which underlie the development and perpetuation of brooding behaviour: 'Birds sit on eggs because certain mechanisms cause them

Table 2.2.   *Three conventional classifications of non-human primate social organization. Compare with Wrangham's (1980) scheme discussed in the text*

Jolly & Plog 1979	Chalmers, 1979	Gouzoules, 1984	Typical mating patterns
Noyau	Solitary species	Solitary	Polygyny/ Promiscuity
Territorial pair	Monogamous family	Family groups	Monogamy
One-male group	One-male (uni-male) groups	One-male units (harems)	Polygyny
Multi-male troop (includes chimpanzees as a sub-category)	Multi-male groups (includes age-graded groups, e.g., gorilla)	One-male units within multi-male groups	Polygyny/ Promiscuity
One-male groups within multi-male troop	Difficult to classify *Papio hamadryas* *Theropithecus gelada* *Pan troglodytes* *Pongo pygmaeus*	Multi-male, multi-female	Polygyny/ Promiscuity/ Polyandry

Note that in most of these broad categories social organization does not afford prediction of a single 'typical' mating pattern.
After Quiatt 1989.

to do so. They are designed to sit on eggs (by natural selection) because the behavior fulfills a function that is important to their survival and reproduction. Questions about both design and mechanism are essential for a full understanding of animal behavior' (MacFarland 1985, p. 3).

It is important to distinguish between mechanism and design because different orders of analysis are involved in examining them (for discussion of the concept of behavioural design in biology see Thompson 1986). Study of behaviour as it occurs, with the focus on individual organisms and their physiological responses to immediate triggering stimuli, affords understanding of the *proximate* circumstances of behaviour. Application of the rules of inheritance and the logic of natural selection to consideration of that same behaviour in context yields explanation of its ultimate origin in genetic adaptation to continuing environmental constraints. The two approaches are complementary, one advances the other, both are necessary to understanding the relation between habitat and behaviour; however, they are like oil and water in that they cannot be shaken up and bottled in a single homogenized theory. This is especially evident when it comes to advancing theories of change. Change in design is the result of selection and adaptation; change in mechanistic response and reaction differences between individuals, while in some sense reflecting genetic adaptation, are more obviously the result of learning. Thus a fully integrated science of behavioural ecology requires study not only of the ultimate origins and proximate circumstances of behaviour but of the developmental and cognitive bases of change and choice.

How do decisions get made?

Behaviour is conditioned by genotype. Genes set the parameters of action; whether some behaviors will occur at all may depend on presence or absence of one or more particular alleles. On the other hand, it is individuals that behave, or perhaps it would be better to say that we tend to organize our observations and to integrate our analyses of system responses and intra-system relations at the level of individual organismic activity. We see with the eyes, hear with the ears, and in all ways encounter our world with the senses of organisms in the diploid phase of their careers. With technical aid we can observe the behaviour of molecules, genes, and chromosomes, see what goes on within and between cells (e.g., the fateful union of haploid progenitors), and record the activities of internal organs; but, marooned in our tight isolation, suffocating sometimes in the separate skins that confine us, it is no wonder if we tend to view behaviour from a particular slant. This bias is perhaps not so remarkable; what is remarkable is the extent to which evolution theorists have managed despite it to picture life from the standpoint of the gene and, from that standpoint, to devise rules which help us to put in evolutionary perspective the behaviour of individuals competing with one another for resources. In the next chapter we will examine the genetic logic which appears to underlie the evolutionary design of primate behaviour. Since individual organisms are not genes it cannot be anticipated that they will act strictly according to the rules of genetic logic, and in Chapter 4 we will ask what other kinds of logic may guide individuals in arriving at decisions, especially decisions as to social action.

Notes

1 This seems as good a place as any, speaking of bias, to comment on our use of personal pronouns. We avoid grammatical oddities such as *(s)he, he/she,* and the plural pronoun *they* applied to single individuals. Occasionally we employ carefully inclusive locutions such as 'he and she' and 'her and him' where the activity at issue may vary in character according to sex – though our usage here is not as consistent as we would like. Most frequently we employ the generic 'he'. This is certainly the easy choice, but it is not an automatic one. Our rationale is that (1) taking the male perspective comes naturally to us; (2) we have grown accustomed to female primatologists using the generic 'she', to the point where it surprises us now to discover one still using 'he'; and (3) division of grammatical labour by sex seems not inappropriate in a discipline in which the sex ratio of practitioners approaches 50/50 (in 1990 the female/male ratio of members of the American Society of Primatologists was 48/52) – equality in economic rewards within the profession is of course another matter.

3   *Emphasizing individual benefits: blind strategies*

1. Introduction
In this chapter we shall examine some general theory which is central to current understanding of behaviour in evolution. Our main concern will be to outline the theoretical underpinnings of current descriptions of non-human primate social behaviour, which we identify with a distinctively modern approach to the study of behaviour in evolution, frequently referred to as *behavioural ecology* or *the ecology of individual behaviour*[1] How those descriptions may inform us as to the evolution of human culture and human societies is the subject of the book as a whole.

2. How did the current view of individual selection emerge?
Socioecology and behavioural ecology
In the late 1950s, Crook (1964, 1965) began an ecologically oriented study of the behaviour of African weavers, birds adapted to a diverse range of habitats. Comparing differences in diet, nesting habits, population density and territorial behaviour against patterns of resource distribution for close to 100 species, Crook found strong correlations between features of habitat and those of social organization (Rubenstein & Wrangham 1986; Crook 1989). Analysis of such correlations, it seemed, might enable one to eliminate those which are more superficial, focusing in on the significant few, and proceed towards understanding how environment could have shaped social organization (Crook 1964, 1965). Crook applied the method to the study of primate as well as avian evolution (Crook & Gartlan 1966; Crook 1970). Other investigations followed (for examples see Sussman 1979), and for a decade or more primate socioecology consisted mainly of a search for meaningful correlations, emphasizing one or another set of variables, between the environment and species-characteristic social systems.

This correlational approach did not demonstrate causal relationships between ecology and behaviour, for reasons which in retrospect seem obvious. The approach was *post hoc* in the sense that it did not relate social system outcomes such as group size or mating patterns to the particular kinds of interactions, intensities, and rates of occurrence which produced them (Rubenstein & Wrangham 1986). Individuals were left out of the account, and no distinction was made between 'random' behavioural variation and variation of adaptive significance (Richard 1981, 1985, and see Sussman 1979). However, to note that early correlational studies were inconclusive is by no means to suggest that they were without consequence. They required data which were ample, rich, and comparable across species; they focused directly on relationships between environment and behaviour (albeit behaviour as reflected in social organizational outcomes); they had as their announced aim evolutionary explanation; they emphasized comparison across large samples of species and representative lifeways; and, finally, they led to the understanding that for animals with complex social systems, for whom complex and subtle acts of cognition and communication were the norm, long-term membership in a particular social group might be the primary ecological relationship between the individual organism and its habitat.

Early socioecology was then, to begin with, largely atheoretical with regard to social behaviour *per se*, but, at the time, neither ecology nor evolutionary biology – the major sources of theory – had much to say about the *behaviour* of individuals. The correlational approach of primate socioecology, centred as it was on species of highly social mammals, provided a background important to the development of genetic logic models of individual cooperation and competition, a background still more important to what might be called the social and cognitive conditioning of cost/benefit investment models as applied to the decision-ruled behaviour of higher vertebrates. Development of these models, relating adaptation of individual behaviour to constraints of setting in the broadest sense, gave rise to a true behavioural ecology. But before that could happen, the relationships between groups and their individual members had to be re-evaluated.

Group versus individual selection

How to recognize or define the loci of natural selection and where to apply measures of relative fitness (i.e., against which collocations of replicating genes) are long-standing, intricate concerns of evolutionary biology. It would be shortsighted to propose that the recent success of sociogenetic theory in addressing certain specific issues of individual cooperation and competition in foraging, predator defence, mate selection, and offspring rearing has resolved more than a few of the problems raised in the debate between group versus individual selectionists. For just one example, the problem of sperm competition, which is almost invariably approached strictly in terms of male competition for females, i.e., as an aspect of genic competition between same-sexed individual diploid organisms, is for many purposes probably better addressed at the haploid level – that is, involving not just competition between spermatozoa but interaction between sperm and egg – and/or in terms of dynamic adaptation (at the diploid level) of organ subsystems (Quiatt and Everett 1982, Eberhard 1985, Small 1988, Sheets-Johnstone 1990). Similarly, punctuated equilibrium in the evolution of species, perhaps especially in the view which has been developed by Eldredge (1985a,b) encourages us to treat species themselves as primary competing units. However, with regard to group selection *per se*, Wynne-Edwards (1962) framed that issue provocatively in terms which, as it turned out, could be nicely clarified and debated at the level of organismic behaviour.

Animal numbers are limited by resources available, nutrient resources most obviously. In the most extreme case, an individual's reproductive success is curtailed either by there being an insufficient local supply of food or by failure to obtain adequate amounts in competition with conspecific others. Twentyfive years ago it seemed to many that in species of social, group-living animals, access to food, and to mates too, could be mediated by group mechanisms such as dominance hierarchies, with some proportion of males relegated to peripheral status or to 'bachelor' groups. There is a general truth in that reflection, but application of such terms to explanation of social organization in evolution tends to confound processes and outcomes of the behaviour of individuals with those of the group as a whole, especially where the genetic consequences of individual migration are not taken sufficiently into account. The term *group mechanism* is loaded with a peculiarly awkward meaning when it is applied to something like observations of status relations among individuals long familiar with one another. A dominance ranking, after all, is not just a graphic simplification of social organization, detachable from a concretely realized whole. It is a summation of some particular set of observed interactions between individual acquaintances. In any group of more than a few individuals, it is unlikely that either a human observer or most animals in the group will have remarked in their entirety the determining features of relationships which are presumed to be expressed by such a ranking, and certainly never all at once. This sort of reification is normal and useful – we intend no pejorative reading of 'reify' – but to ascribe to an analytic concept like 'status hierarchy' the properties of a *mechanism*, capable as an entity in itself of influencing the actions of those individuals whose relationships it was devised to describe, invites confusion. (See in this connection the more general discussion by Haraway (1978) of how concepts central to 'animal sociology' may be permeated and thus subverted by other concepts which are inappropriate and unrecognized.) Wynne-Edwards, in *Animal Dispersion in Relation to Social Behaviour* (1962), treated the social behaviour of animals in groups in just this way. Certain mechanisms of group social behaviour were supposed to provide an early warning of imbalance between population numbers and resources; some of those mechanisms also induced corrective behaviour, enabling groups, populations, and species to regulate numbers comfortably within a relatively narrow range of variation – that is, without the boom and bust swings that Wynne-Edwards assumed must occur if environmental constraints on individual reproduction were not mediated by group mechanisms. In Wynne-Edwards' view, ritualization of aggression also helped to ensure that competition within groups would resolve problems of incipient scarcity without the unreasonable cost that selfish, diehard struggle between individuals must, it seemed, entail. Thus, a group characterized by good internal communication, high levels of cooperation, and effective ritualization of aggression would have advantages relative to groups in which such conditions did not obtain.

We will not rehearse Wynne-Edwards' arguments in detail, nor will we recapitulate all of the objections raised against them. To review fully the issues which engaged proponents of individual selection and advocates of group selection in the debate which followed publication of *Animal Dispersion* would take a book in itself – a book which, we hardly need add, has been written in its several versions by many of the debaters themselves (see e.g. Williams 1966 and Maynard Smith 1975) and not a few of their first audience. The achievement of Wynne-Edwards lay, first, in having outlined a number of highly specific means by which the reproductive behaviour of individuals appeared to be regulated toward the interest of the group as a whole, and second, in having defined the interest of the group, not in the usual vague terms of general benefit, functional efficiency, and economy of the whole but explicitly as an increase in numbers relative to those of competing groups, against a background of limited resources. He was, to be sure, not consistently explicit about the level of organization at which groups should be described and compared, about the character and permeability of those groups' boundaries, or, as we have intimated, about the consequences of individual migration in and out of competing groups to the collective genotypes of those groups, where genotype can be assumed to have a bearing on competitive behaviour.

A major obstacle to acceptance of Wynne-Edwards' theory of how animal numbers were regulated was that David Lack's formulation (1954) of the same phenomenon, framed in terms of individual selection, provided a simpler and to most evolutionary biologists a more convincing explanation. Also, following Watson and Crick's exposition (1953) of the physical structure of DNA, much new information had been developed on the function of genes in life, and experimental geneticists were busier than ever constructing controlled studies of natural selection in laboratory populations. These studies, while they may have had little to say about the richness and complexity of ecological relationships in nature, furthered the development and application to living organisms of reliable measures of individual fitness.

Kin altruism

Theories of group selection tend to be tied more closely to system and economic exchange theories and to ideas of organizational savings than to genetics; Wynne-Edwards himself has suggested that his grasp of the genetic basis of evolution must seem insecure to 'a laboratory geneticist' (Wynne-Edwards 1964). Theories of individual selection put the emphasis squarely on differential replication of genes. To individual selectionists, the key concept in any theory of behavioural evolution was reproductive maximization by individuals and consequent relative increase in individual fitness. As Hamilton emphasized (1964), following an earlier casual observation of Haldane, energetic investments in fitness at the level of the gene need not be confined to an organism's own offspring and direct descendants, they can be conferred on any reproducing individual carrying genes identical with those of interest to the investor, with reproductive profit in direct ratio to proportion of genes shared, or (the same thing put differently) to the likelihood that a gene for any trait of interest will be identical in both parties. That likelihood could be calculated for biological kin, collateral or direct, according to degree of relationship. Assuming that animals are capable of recognizing behaviourally close collateral kin as well as their own offspring,[2] natural selection might well encourage energetic investment in the well-being and reproductive success of those kin – to the extent warranted by the proportion of genes shared in common, offset by the costs to the animal's own direct reproductive success. To the extent that a group consists of individuals who are, on average, closer kin than are those in adjacent, competitor groups, kin altruism and inclusive fitness (Hamilton 1964, or *kin selection*, as it was termed by Maynard Smith 1964) would go far towards explaining the apparent contradiction involved in a sacrifice of individual fitness to the 'greater good' of the group. Gains in inclusive fitness by more distant relatives can be assumed to offset reduction in that portion of fitness contributed by direct descendent genes. While the logic of kin selection seems impeccable, it is important that we do not ignore the very real problem of measuring fitness in real populations. Another major problem too seldom remarked upon has to do with time depth. It is clear that fitness must be measured in production of descendants (direct or lateral) who are themselves reproductively fit; but what counts in the end is long-term success of a lineage and how far down the generational trail we can expect to extend our speculations about the 'adaptiveness' of a particular behaviour – and, of course, as palaeontologists insist on reminding us, all lineages go extinct with the species.

Reciprocal altruism

What about group associates or neighbours to whom an animal is *not* closely related? The workings of kin altruism have been most satisfyingly demonstrated in social insects with well defined, nest-bound lineages and internal social castes in which distribution of genotypes is highly predictable from one generation to the next, though there appears to be a surprising amount of individual variation within colonies of social insects (Robinson & Page 1988; Frumhoff & Baker 1988). In most vertebrate species, boundaries between lineages or, indeed, between reproductive groupings larger than the mated pair or cooperatively breeding family are not neatly defined. Although membership may be relatively stable in a territorial family of gibbons, a troop of baboons, or a community of chimpanzees, (or a pack of wolves or a cooperatively breeding family of nutcracker jays), in every generation members of one or the other sex or of both sexes emigrate from the group (Table 3.1). This has implications both for inbreeding avoidance and for cooperation. The influence of migration patterns on population density, on gene dispersal, and on corporate genetic investment are not easy to ascertain; the few instances in which kin altruism has been shown to operate in vertebrates have involved close kin living in close proximity to one another (e.g., naked mole rats (Jarvis 1981) and California ground squirrels (Sherman 1981).

Kin altruism, it seemed, could explain much but not all self-denying cooperative behaviour. What about those instances in which an animal behaves in such a way as to foster reproductive fitness in another with no inclusive fitness benefits in return? Trivers (1971) proposed that under certain conditions selection should work to favour reciprocal investments in fitness even though the animals involved were so distantly related to

Table 3.1. *Tenure in group by sex, in social mammals where male transfer predominates*

Species	Mean female age at first conception (months)	Mean male residence (months)	Does female age at first breeding exceed average male residence in a single group?
One-male groups or territories			
Yellowbellied marmot (*Marmota flaviventris*)	36	27 R	Yes
Blacktailed prairie dog (*Cynomys ludovicianus*)	24	<24 R	Yes
Red deer (*Cervus elaphus*)	40	36 R	Yes
Wedge-capped capuchin (*Cebus olivaceus*)	72	120 R	No
Redtail monkey (*Cercopithecus ascanius*)	24	<24 R	Yes
Campbell's guenon (*C. campbelli*)	36	<34 R	Yes
Blue monkey (*C. mitis*)	66	<40 R	Yes
Gelada baboon (*Theropithecus gelada*)	54	43 R	Yes
Grey langur (*Presbytis entellus*)	36	28 R	Yes
Purple leaf monkey (*P. senex*)	48	36 R	Yes
Multi-male groups			
Eastern grey kangaroo (*Macropus giganteus*)	24	12 *α*	Yes
African lion (*Panthera leo*)	38	26 R	Yes
Ringtail lemur (*Lemur catta*)	30	<24 R	Probably
Vervet monkey (*Cercopithecus aethiops*)	33	32 R	Yes
Yellow baboon (*Papio cynocephalus*)	72	<51 R	Yes
Rhesus macaque (*Macaca mulatta*)	66	19 R	Yes
Japanese macaque (*M. fuscata*)	54	<30 R	Yes
Torque macaque (*M. sinica*)	60	50 R	Yes

For mean male residence: R, residence; α, tenure of α rank. Field data were used to estimate the age of females at first conception wherever possible. Average male residence was the estimated duration of reproductive activity in males within a particular group. In the majority of species, this was equivalent to mean residence per group in non-natal breeding groups but for grey kangaroos the mean tenure of α rank was used and for red deer the one another that kin altruism could be ruled out. Chances of selection for reciprocal altruism, according to Trivers, were

> greatest (1) when there are many such altruistic situations in the lifetime of the altruists, (2) when a given altruist repeatedly interacts with the same small set of individuals, and (3) when pairs of altruists are exposed 'symmetrically' to altruistic situations, that is, in such a way that the two are able to render roughly equivalent benefits to each other at roughly equivalent costs.

Preconditions postulated by Trivers for the evolution of *reciprocal altruism* include: 'long lifespan; low dispersal rate; life in small, mutually dependent, stable, social groups . . . ; and a long period of parental care' (1971). These can also be viewed as a set of preconditions for limited and gradual dissemination of genes, raising the question whether Trivers' theory of reciprocal altruism should be regarded as *alternative* to Hamilton's model or as *ancillary*, specifying the proximate variables that influence kin altruism in its operation in real populations (Quiatt 1986b). If, as has been suggested, kin *recognition* in mammals is primarily a function of spatial propinquity and a history of friendly interaction (Bekoff 1981; Holmes & Sherman 1982, 1983), there may be little point in recognizing conspecific reciprocal altruism as a behavioural phenomenon distinct from kin altruism. It has been argued that, for primates, friendship supersedes kinship (Frederickson & Sackett 1984; Quiatt 1986b, 1988a), but perhaps the real question at this level of analysis is how to distinguish behaviourally between friendship and kinship. This question may be resolved for non-human primates as studies of social cognition achieve a level of sophistication equal to those of kin recognition *per se* (Fletcher & Michener 1987). At present, however, we find it difficult to understand how, in calculating the effects of kin selection, we are to eliminate effects which are due to kin altruism *per se*, and vice versa.

Parental investment and parent–offspring competition
Strategies of female and male parents
A basic theme of socioecology is that animal numbers depend on resources available, balanced against predation pressure, and that the spacing of animals in a given species is largely determined by the pattern of resource distribution coupled with the character of predation (Crook 1964; Emlen & Oring 1977; Slobodchikoff 1984; see Barlow 1988 for review). In animal mating theory, mating systems are viewed as adaptive to these patterns, and it follows that understanding of social systems in evolution must take all such factors into account.

But we must also take into account fundamental differences between the sexes, beginning with an unequal investment in production of gametes (anisogamy). This inequality has important implications for mating strategies (Williams 1966, and see our discussion in the previous chapter of Wrangham's theory concerning female-bonded species of primates). Females produce relatively few, relatively large gametes during their life-time and in many species are less mobile than males, who produce smaller gametes in greater number. In this situation, and given that fertilization of an egg normally involves its union with a single male gamete (though we must note that that is not quite the 'given' that it once seemed to be), it would appear that females are a *limiting resource* for males, whose reproductive success relative to other males depends at the outset on how many eggs they can fertilize. Females, so the numbers suggest, need not strive to monopolize a male partner's spermatozoa, except insofar as guarding against fertile inseminations outside the relationship may help insure that parental care is not likewise diverted (Small 1988). They should however select among males competing for them – picking and choosing on the basis of traits likely to enhance their reproductive success, e.g., genetic or behavioural compatibility and whatever contributes to the quality of male parental care, beginning in many species with ability to defend territory (Trivers 1972; Halliday 1983). A few moments' reflection on all that is involved in female choice, and the many ways in which female primates may exercise choice at one or another juncture throughout each reproductive cycle, should convince the reader that where primates are concerned 'passive' is not an appropriate modifier of female 'choice', that the selection and often quite active search that is involved creates plenty of room for competition with other females. This observation of course is intended as a caution against over-generalization, not as refutation of the anisogamy argument.

Trivers (1972, 1974) generalized the argument and extended examination of differences in male and female strategies of parental investment over the course of the parent-offspring relationship. Parental investment began with metabolic investment in production of gametes but included also energy expended in gestation and lactation and, subsequently, in feeding and protecting young until they were established as reproductive individuals in their own right:

> The time sequence of parental investment analyzed by sex is an important parameter affecting species in which both sexes invest considerable parental care: the individual initially investing more (usually the female) is vulnerable to desertion. On the other hand, in species with internal fertilization and strong male parental investment, the male is

always vulnerable to cuckoldry. Such vulnerability has led to the evolution of adaptation to decrease the vulnerability and to counter-adaptations. (Trivers 1972)

Reproductive behaviour – courtship, mating and parenting – involves relationships that are characterized by both cooperation and competition. The reproductive goals of female and male partners are parallel but by no means identical. Trivers emphasized (some would say over-emphasized) the competition inherent in the mating/parenting relationships. To Trivers this competition, and associated differences in female and male investment in offspring, appeared to provide the key to understanding sexual selection and the evolution of sexual strategies.

Parental investment in female and male offspring

Where polygyny obtains, as it does in the great majority of mammalian species, variance in reproductive success is likely to be greater among males than among females, and parents who can afford to invest heavily in either sex should invest more heavily in males. Those who do will be survived by greater numbers of grandoffspring than those who invest equally in offspring of both sexes. Thus runs the argument proposed by Trivers & Willard (1973), and it is supported by long-term studies of reproductive success in wild populations (Le Boeuf 1974; Clutton-Brock *et al.* 1979) – at least where pre-weaning investment is concerned:

> In several species where contests involve grappling, pushing or charging, there is evidence that reproductive success among males is related to body size . . . which is usually well correlated with early growth rates, maternal milk yields and maternal body condition . . . In such cases, it seems likely that a male's reproductive success is strongly influenced by maternal investment before weaning and parents should invest more heavily in their sons than their daughters during this stage. (Clutton-Brock & Albon 1982:224)

However, post-weaning life histories of male and female offspring may call for different parental investment strategies. In many terrestrial mammals, not least in primates, daughters remain in their natal group or adopt home ranges adjacent to mothers, while sons disperse. In these circumstances, mothers are likely to continue investing in daughters long after weaning, in many cases well into and even throughout reproductive careers. Of course, if daughters were ejected by mothers, resources saved might be lost to unrelated animals, and the cost of a daughter's continued presence may be more than offset by benefits accrued in defence against prcdation, defending food patches against conspecifics, and allomaternal help (Clutton-Brock & Albon 1982).

As these observations may suggest, it is difficult to say where to locate the end-gate, for purposes of analysis, in the history of a given parent–offspring relationship. At what point if at all might the parental investment model be dispensed with and analysis of costs and benefits to both individuals be reinitiated from the broader perspective of kin altruism? The problem perhaps is not so much locating the end-gate as making gate-keeping decisions: what are the problems of particular interest to the research underway, and to what kinds of information will the researcher therefore want to attend?

A similar question can be asked about the starting gate, which is equally though not as obviously arbitrary. Parental investment begins with the anisogamic premise that female investment at the moment of conception is greater because the egg is larger than the sperm, but do males really invest but one sperm in fertilizing a zygote? Fedigan (1982) has argued to the contrary, that

> [T]hey invest millions in a copulation, and countless millions in the several copulations which a male usually undertakes with one female before a conception results. More importantly, males invest significantly in terms of energy and risk to future fitness in order to reach the point of copulating with females (p.302)

Is Fedigan's criticism legitimate? Apart from the fact that Trivers ruled out, in his definition of parental investment, energy expended in seeking out, competing for, and courting prospective mates, Fedigan's observations appear to embody the same flaw that has frequently been remarked (by, e.g., Boucher 1977; Maynard Smith 1977; Dawkins 1979; and by Fedigan herself, 1982) in the parental investment model as originally formulated. Parents (and other cost/benefit strategists) should attend to imminent costs and future possibilities, not to irretrievable investments of the past. They should not behave like some poker players, who out of concern for protecting their ante ignore the fact that the odds change with each draw, the cost of losing with each raise in the bet.

Parental investment theory has great explanatory power. Nevertheless, there remains something discomfiting about a theory which seems to take for granted that there will be no insurmountable problems in the way of totting up and comparing parental investments by sex. As Barlow (1988) has pointed out, while it may be 'relatively easy, *and intuitive*, to establish which sex invests the most' in offspring, it is nevertheless 'difficult if not impossible to . . . [arrive at] a general construct because there is no common quantitative currency that applies to both sexes' (p.56, emphasis added). According to Maynard Smith (1984), 'The real difficulty in analyzing parental behaviour arises . . . because the optimal behaviour for one parent depends on what the other parent is doing.' Maynard Smith suggests that game theory be applied to discover 'a pair of strategies . . . which together form an evolutionarily stable strategy or ESS . . . , in the sense that it would not pay a male to depart from strategy M so long as females continue to adopt F, and it would not pay a female to depart from strategy F so long as males continue to adopt M' (p.218).

3. How has the modern view of selection for behaviour influenced our understanding of evolution and evolutionary biology?

The twin pillars of the modern theory of social behaviour are the principles of individualistic reproductive maximization (G. C. Williams 1966) and kin selection (W. D. Hamilton 1964). With the development of these principles, animals came to be viewed as individuals who were armed with many behavioral options in their struggle for maximizing either their own reproduction or that of their relatives. (Rubenstein & Wrangham 1986:4).

Behaviour can provide the context for selection resulting in anatomical change. Expressed in terms of a 'priority' in evolution of behaviour over morphology, this idea has provided something like a first principle of palaeoanthropological speculation about hominid origins – while at the same time raising awkward chicken-and-egg questions, so that palaeontologists as a rule have wisely referred discussion of change in behaviour to dated change in morphology. However, the current emphasis on social behaviour, on social 'struggle', and on the exercise of behavioural options by individuals in competition with other individuals for local resources is something new as a serious and sophisticated theoretical concern; it has changed the way which anthropologists, primatologists, and biologists in general approach the study of evolution and evolutionary process. The individual, hitherto left standing in the wings or altogether off the evolutionary stage, has moved front and centre and been assigned something like cast credit as the agent in which determinative behaviour is organized and manifested.

The new emphasis on individual players has changed the way we think about natural selection, genetic models of adaptation, and the ecological contexts of evolutionary change – and is beginning to change our ideas about society and culture and about the study of animal and, especially, non-human primate behaviour. Anthropologists are far more comfortable these days than they had been in even the recent past (prior to, say, 1980) discussing the behaviour of monkeys and apes in terms of social strategies, tactical decisions, and the implementation of cost-accounted decisions in thoughtful action (see Standen & Foley 1989a; also Whiten & Byrne 1988b for review and critical commentary of deception as it pertains to intentional action, an issue we will discuss more thoroughly in Chapter 8.)

Natural selection and evolution

The current view does not involve a revision of basic theory. Rather, it constitutes a clarification and refinement of Darwin's ideas concerning natural selection, bringing them into closer accommodation with current knowledge of the gene, its physical character and physiological function, and with our present understanding, conditioned by a theory-based science of ethology, of how genotype may constrain differences in the behaviour of individuals toward kin and non-kin and, indeed, of social behaviour in general. Williams's representation of the individual selectionist position (1966) shows how thoroughly that position is founded in Darwin's theory of natural selection. Sociobiological explanations of behaviour are not just compatible with Darwin's ideas, they are implicit in them, a natural extension of Darwinian theory. The main extension involves, first, explicit recognition that homologous alleles are invariant, physically indistinguishable whether alike by direct descent, close collateral descent, more distant replication, or mutation. A second involves understanding, equally clearly, of the main implication of that recognition – how, as spelled out in detail by Hamilton (1964), we may broaden the definition of individual reproductive fitness along collateral lines of genetic investment.

In sociobiology, it should be noted, emphasis is placed on 1. natural selection as the mechanism primarily responsible for evolutionary change (implicit in the twin assumptions, guiding principles almost, that behaviours are genetically influenced and that they are adaptive), and 2. alternative social behaviours as traits under selection – although this is very much an oversimplification. Sociobiology's emphasis on selection and focus on individuals in social competition smacked of Social Darwinism to many social scientists. Overenthusiastic use of dramatic metaphor by some sociobiologists, along with a certain recklessness in the deployment of cost/benefit decision and investment models borrowed from economists – 'recovered' may be the better term, given the influence that accounts of animal behaviour had exercised on economic theory beginning with Malthus (1798) – did little to correct this impression. In the middle and late 1970s political debate on sociobiology flourished; during that period, biologists and social scientists who did not select with care the sessions they wished to attend at professional congresses risked subjecting themselves to mind-numbing confrontations in which colleagues arguing at cross purposes rarely managed to address the same issues.

Anthropological genetics

Ever since Mendel, formal genetics has taken as its primary concerns establishing the heritability of particular traits and determining modes of inheritance. At this first level of analysis, concern for individual differences and for the specific distribution of traits among kin – i.e., in a kindred composed of individuals whose several relationships to one another have been experimentally determined or can be verified – is clearly central to the study of hereditary systems. We do not propose to put forward here even a capsule history of genetics, only to remind readers that this particularistic approach did not in itself contribute much toward understanding how such systems evolve. However, a shift of analytic focus from the expression of a trait in an individual, and the distribution of a family of traits in a kindred, to genetic loci *per se* and to relative frequencies population-wide of alleles at one or another locus, allowed population geneticists of the classical period to define evolution as *change in gene frequency* – an operational definition which proved to be as powerful as it was simple. The theory on which this definition was based derives from Hardy's (1908) observations concerning the constancy of gene and genotype frequencies under certain ideal conditions. These conditions ruled out change in gene frequencies resulting from mutation, selection, migration and drift (equivalent to sampling error in generational replacement, as population increase through births is treated in the models of population genetics). Geneticists were encouraged to think of evolution at the population level in terms of just these specific mechanisms of genetic change, by definition a complete set, and a suitable basis was provided for quantitative analysis of the dynamics of that change, albeit limited to the short term and to one or a few families of alleles at known loci.

The definition of evolution as change in gene frequency, the specific mechanisms of change derived from that definition, and the foundation which these establish for quantitative analysis of microevolutionary processes were major contributions to evolutionary theory. They made up the backbone of the New or Modern Synthesis, and they changed the character of anthropological research into human variation. Anthropological genetics, when the phrase is not used synonymously with human genetics or medical genetics, usually refers to population genetics as developed in the classical period (i.e., approximately the first half of the present century but especially that period from the early 1930s through to the 1950s) and applied to anthropological concerns. Contrary to what might have been anticipated, given its relatively restricted temporal focus and its attention both to the role of migration and to that of small populations in human evolution (*drift* is sometimes referred to, in terms of results, as the *small population effect*), population genetics did not at first infuse ethnology and sociology with fresh concepts as it did biological anthropology, or with models for exploring the dynamics of competition for local resources; nor did it effect a markedly closer integration of research in those disciplines with that in biological anthropology. There seem to be two reasons for this. One is that the individual, while of considerable importance to social scientists and perhaps especially to ethnologists, appears in the models of population geneticists as little more than an obligatory abstraction intermediate between the genetic locus and the gene pool. The other is that, despite the deterministic function that natural selection exercises as a mechanism in those models, randomness is built into them too as a necessary feature central to their construction. In the construction of genetic models, randomness of course is not exactly equivalent to 'chance' in nature; in models of natural systems it is introduced where necessary as a kind of replacement analogue for all that we cannot know about the workings of nature. It is hard to see how, without randomizing mechanisms, a model of a complex system, which is dynamic and with a temporal dimension, could further our understanding of process. However, the result of leaving out individuals and introducing randomness at critical stages in the operation of a population genetics model is to make the model inappropriate for the study of social and cultural behaviour. It can be argued that the classic models of population genetics were never intended to illuminate behavioural mechanisms, but this remains a weakness, for the complexities of social and cultural behaviour need to be included in models of the genetic evolution of populations. Failing that, the study of population structure must remain the study of inbreeding in small populations and of assortative mating for physical traits assumed to be wholly or largely determined by genetic inheritance (as, e.g., red hair, light skin coloration, height).

To the extent that anthropological genetics has dealt with the behaving individual as something other than a medical case, it has tended to treat individuals as organic billiard balls or gas molecules. Until very recently, little attention was paid to learned, culturally constrained individual behaviour as a determinative force in evolution. Neither, it might be argued, do the new evolutionary models of natural and cultural selection for particular, specified kinds of behaviour, in competition between and among individuals, put much emphasis on the learned bases of behaviour or on behaviour itself as determinative. But some do (e.g., Boyd & Richerson 1985; Plotkin & Odling-Smee 1981; see also Harrison 1988 and Dunbar 1989), and all of them reflect a world in which, by implication at least, the behavioural strategies of diverse individuals capable of learning constitute an important element of the social interactional setting, hence, in the current view, of the natural environment in which selection is assumed to operate. Therein lies a significant difference from older models, and it has gradually brought attention to bear on the experiential and cognitive bases of choice in consideration of alternative behaviours.

4. How has behavioural ecology influenced the study of primate behaviour?

The approach we have been discussing derives from a re-examination by evolutionary biologists, ecologists, ethologists, and geneticists of natural selection operating at the level of individual interaction. Premises guiding recent and current studies of social behaviour are 'that the starting point for analysis is the individual, not the group or species to which it belongs . . . [and that] the *adaptive significance* of a behaviour must be understood in the context of reproductive success: how did the behaviour contribute to individual fitness?' (McKenna, 1982, p.56, author's emphasis).

There are a number of implications here for the study of animal behaviour in general. They concern various aspects of study, including, for example, methodological focus on dyadic interactions, competition versus cooperation, cost/benefit decision-making, models of evolution versus models of learning, and field versus laboratory studies. We will suggest for each of these what some of the implications are.

Dyadic interactions

Description of behaviour necessarily entails recording what individuals do, that of social behaviour what they do relative to one another. The individual selectionist argument provides a theoretical rationale for assigning priority in description and analysis to dyadic interactions in groups. A good example is Hinde's emphasis, frequently reiterated, on description of individual relationships as central to the analysis of primate social organization (Hinde & Stevenson-Hinde 1976; Hinde 1983a, 1987). For those whose inclination is toward observing behaviour in the field, the attraction of models which are primarily descriptive may be particularly strong. Hinde's approach is of interest on both these counts; we shall discuss it in detail in the next chapter.

Competition versus cooperation

Concern for the competitive aspects of animal behaviour is hardly new. Scientists interested in competition between groups have tended to treat outcomes of that competition as reflecting differences in the social organization of competing groups, with the analytic focus on features of social organization deemed to be relatively advantageous or disadvantageous (as in, e.g., Wynne-Edwards 1962). Individual selectionists, on the other hand, have tended to view social organizational structures as emergent and dependent on the character of competitive interactions between individuals. Emphasizing the primacy of individual–individual competition, they have focused analysis on alternative strategies available to individuals in specified contexts of competitive behaviour (as, e.g., G.A. Parker 1974; Dawkins 1980; Dunbar 1984). An important goal for future research is to obtain reliable accounts of the precise nature and outcome of such competition; this is best achieved where what is at stake is local resources which are essential to reproductive success, making a difference in ways which are clearly definable and, ideally, measurable.

Kin and reciprocal altruism do not so much temper individual competition as extend it – they are means by which the individual organism enlists relatives and acquaintances in his or her personal game, as suggested by the saying 'I against my brother; I and my brother against our cousin; I and my brother and our cousin against . . .' (Chatwin 1987; see also Hughes 1988). Competition may be more evident in some contexts of interaction than in others, but from the individual selectionist standpoint all behaviour is coloured by competition. A closely cooperative partner in social behaviour may be simultaneously an opponent with regard to some aspect of reproductive energetics, as in a mating pair relationship. Since opponents/partners are often characterized by asymmetries of sex, developmental age, and cohort status, with characteristic differences in competitive strategies, the forms of competition are similarly complex and varied. Still, the ultimate measure of success, explicit in the second of McKenna's premises (see Chapter 2), is single and clear-cut; individual selectionist models of competition gain power from their ability, whatever the area of concern, to focus attention 'on a limited number of measurable variables that ultimately are associated with reproductive energetics' (Rubenstein & Wrangham 1986:5 – the reference here is specifically to Maynard Smith's (1977) application of game theory modelling to optimal mating strategies).

Cost/benefit decisions

The basic sociobiological model of behaviour in evolution is one of change – underlying gene structure altered by selection favouring one or another alternative strategy of individual competition – or of balance reflected in an evolutionarily stable strategy (ESS) of two or more behavioural alternatives which are maintained by competing selective forces (Maynard Smith 1974, 1976, Dawkins 1980). In either case it is an economic model, with the value residing in reproductive fitness being measured ultimately in the contributions by individuals to the population gene pool. Offspring (and other close relatives in the next descendent generation, but especially offspring) constitute a basic currency against which other currencies, more useful for on-the-spot comparison of the effectiveness of alternative strategies (i.e., constant units employed in the construction of time/energy budgets or in estimation of rates and frequencies of behaviour) can be referred. The basic model, being general and abstract, does not specify the character and extent of genetic determination of behaviour, and it treats individual cognitive review of situational variables (assuming that such a review should occur) as being outside the analysis. Whether or not individuals reflect on the advantages of one strategy over another is beside the point, that they act 'as if' they were choosing to exercise a particular strategy is enough (cf. what David Barash (1977:63) has called the 'Central Theorem of Sociobiology, a fundamental hypothesis . . . [which] states: When any behaviour under study reflects some component of genotype, animals should behave so as to maximize their inclusive fitness.').

Sociobiological theory tends to leave unresolved the question whether, over time, decisions as to which strategies of behaviour will prevail come to be predetermined (through genetic adaptation) for particular ecological contexts, or whether they are recalculated in experience by individuals of every succeeding generation. If the former, then we return to the question of who the players are in this evolutionary game; if the latter, then we must wonder why it is that we ourselves do not follow the rules on a more conscious level of play than we appear to do (Hughes 1988). If a behaviour has the function of maximizing inclusive fitness, and if that function is apparent to the external observer, then human beings, who

LIVERPOOL
JOHN MOORES UNIVERSITY
AVRIL ROBARTS LRC
TEL. 0151 231 4022

are supposedly capable of objectifying their own behaviour, should be able to identify fitness-maximizing functions and make the most of them. However, as has often been remarked, the goal of human behaviour as assessed by external observers is not always and perhaps not often the goal that the individual observed claims to have in mind. For human beings and other animals, the goal that an individual has in mind may differ from the one that he wishes external observers to think that he has in mind. Thus behaviour may be 'designed' to deceive others. It has often been asserted that the best deceivers are those capable to some degree of deceiving themselves, and it may be that self-deception has played a role in the evolution of human cognition (Lockard & Paulhus 1988; Trivers 1985).

Human behaviour apart, that is, even considering the less problematic application of cost/benefit accounting to the behaviour of non-human animals, the cost/benefit accounting approach to evaluation of the consequences of behaviour strategies so far has proved most useful for formulating hypotheses and 'pre-testing' them in the field against real life outcomes. It is clear that cost/benefit accounting has been very useful to sociobiologists and socioecologists. What is at present not so clear is whether and how animals may themselves practise cost/benefit accounting, i.e., whether the phrase can be read as something more than layered metaphor for all that remains to be discovered about the relation between blind strategies and tactical decisions. References to cost/benefit accounting, where a literal reading is intended, too often still signal conflation of concepts appropriate to genetic adaptation on the one hand and learning on the other.

Evolution versus learning

Sociobiology explicitly aims at explaining the ultimate origins of social behaviour. Although it can hardly be said that primatologists have ever been *un*interested in the evolutionary origins of behaviour, early emphases where social behaviour is concerned were on individual learning (e.g., of dominance or mothering roles), socialization and enculturation (whether of immigrants or of infants born into the group), and acquisition of new habits such as sweet potato washing and wheat rinsing by Japanese monkeys (Kawamura 1959, Kawai 1965). To the extent that anthropologists in particular may have tended to apply to non-human primate groups a synchronic, structural–functionalist model of social organization (as in the early socio-spatial models of Old World monkey group organization, e.g., Imanishi 1960; Hall & DeVore 1965; for review see Quiatt *et al.* 1981), they presumably would have been less interested in origins than in outcomes – more interested, that is, in the contribution of learning and development to individual performance of roles essential to normal functioning of the group as a whole, under particular environmental circumstances.

Disciplinary affiliations apart, until recent years analyses of proximate constraints on behaviour were by no means consistently unified by Darwinist assumptions. Where they were not tied explicitly to models of individual learning, they often have rested on learning models by default. In our view, selection is the concept which best illuminates connections between behavioural lability and the proximate circumstances associated with variable habitats. Analyses of behaviour which take this into consideration are better able to inform us as to how both capacity for learning and learning processes themselves have evolved. It seems not too strong an assertion to say that current studies of primate social behaviour in natural habitats, insofar as they are informed by principles of sociobiology *and* socioecology, reflect a revitalization of behavioural primatology.

Field versus laboratory studies

Studies of behaviour in nature and in laboratory settings have differed in method and, perhaps less obviously, in underlying theory – though consideration of those differences must take into account too differences between European ethology and North American comparative psychology. Studies in the latter tradition have been laboratory- and learning-oriented, focused for the first half of the century on a few species, predominately the Norway rat (Beach 1950), and 'on partsystems' (e.g., concept formation, the mechanisms of hunger, and the influence of hormones on sexual performance) abstracted from the whole behaviour of an organism (Mason 1968). In the former, by contrast, studies have been comparative and informed by evolutionary theory – though, to begin with at least, not always by insights into the character of individual variation. The same criticism can be made of sociobiology as developed in the 1970s, but we seem to be growing wary at last of references to 'species-specific' and 'species-characteristic' behaviour (see Standen & Foley 1989a); similar wariness is advisable in the case of reference to habitat types, for habitats may vary considerably from deme to deme within a species.

Ethologists were by no means strangers to the laboratory, and controlled experiments were basic to the 'classical' ethology of Konrad Lorenz and Niko Tinbergen. But in North America, and prior to the rapprochement of comparative psychology with European ethology beginning in the 1950s – in which, it should be noted, anthropologists and developmental psychologists figured prominently (most notably perhaps in a key series of interdisciplinary symposia sponsored by the Macy foundation: Conference on Group Processes 1954–58) – students of animal behaviour in nature tended to work out of a tradition of natural history, inspired perhaps as much by Ernest Thompson Seton as by Darwin.

We do not wish to exaggerate the significance of this rather obvious, frequently remarked dichotomy between, on the one hand, evolutionary theory and an interest in behaviour as a whole-system activity, pursued primarily via study of animals in naturalistic habitats, and, on the other hand, learning theory, an interest in part-systems, and study of behaviour subject to close experimental control. Field studies of *primate* behaviour, getting off to a late start, have always been informed by what we have learned (and mislearned) from observation of animals in laboratories and zoos, and at the same time field investigations have unearthed questions for further laboratory research. More important from this historical perspective, long-term field studies have provided a facilitative context *in* the field for 'natural' and controlled experiments dealing with the adaptive character of particular behaviours (e.g. Kummer 1968; Rowell 1967). This approach is supported by the modern view of social behaviour, for if social behaviour is the adaptive consequence of selection for particular alternative strategies exercised by individuals in competition with one another, understanding of ultimate origins can best be gained by studying behaviour in environments which are most similar to those in which selection is presumed to have operated. Of course, we cannot ever know very precisely what those palaeo-environments may have been like with regard to socioecological components, but that is another issue.

Ecology as an interest in relations between species in nature has been around for a long time, but the development of ecology as a modern discipline is something else. We leave it to ecologists to date the inception of this development, but surely it cannot be dated much before 1960. Prior to that time, 'most applications of ecological principles . . . pertained to the management or control of specific resources or species' (Odum, 1971:405). It may be that, as Odum goes on to suggest, with increasing application of those principles to holistic studies of overall productivity, food chains, energy cycles, etc., practice caught up with theory, but practice also contributed to theory. A broader systems approach furthered the realization that supply and living space functions of the environment are 'interrelated, mutually restrictive, and . . . limited in capacity' (p.405), a realization which quite apart from signalling an attitude change with regard to human use of the environment (see also Odum 1969 & 1970) seemed to call for an animal ecology thoroughly infused with behaviour theory.

5. How has primatology influenced behavioural ecology?

Modern field-oriented primatology and modern behavioural ecology have been closely contemporaneous in their development. Concepts and hypotheses put forward by ecologists have been a tremendous stimulus to primate field studies: examples are the recognition of *K* selection principles (MacArthur 1962), viewed as of particular importance in primate evolution (Lancaster 1984), the model for evolution of polygamy derived from ecological studies of birds (Orians 1969), and Maynard Smith's application of game theory to a model of mating system energetics (1977). Ecological concern for the environmental determinants of mating system variability has emerged as a kind of basic strategy, or first step, in the study of social systems in general. But it is important not to overlook contributions in the other direction. Field primatologists emphasized almost from the beginning the importance of studying the same group over all seasons of the year, of conducting multi-year studies or restudying the same group periodically to ascertain longer cyclic patterns of change, of comparing groups of the same species living in different habitats, and, in studies across species, of using similar techniques and terms of description so as to further comparative analysis. Following the practice of Japanese primatologists, individual animals were identified for the record. At some sites (e.g., Cayo Santiago off Puerto Rico), animals were physically marked; at others, continuity of personnel and in some cases the pronounced character of individual variability sufficed – as with the chimpanzees of the Gombe Stream Reserve (Goodall 1965, 1971, 1986a,b) and the Tocque macaques studied by Dittus (1980). Data thus accumulated, especially data on genealogical and other associations within and between groups, provided a record of social life unparalleled in detail and, more important, in what that detail revealed about the ecological significance of the group to its individual members.

Data from primate studies were, from this standpoint, important to the development of a behavioural ecology focused not just on the adaptation of a given species to coexistence with larger communities of species but on the ecosystem as a whole, beginning with interactions between individual organisms (see Standen & Foley 1989a, but cf. Richard 1985:35). Also, from around 1960, studies of primate behaviour (in the United States at least) were well-funded, and primatology began to attract great numbers of students from a variety of disciplines, especially perhaps, to begin with, from anthropology. Primates quickly became the most studied of mammalian vertebrates, and an important focus of that study was primate social behaviour. Anthropomorphic anticipations and interpretations apart, the social behaviour of non-human primates was clearly interesting and highly variable both across and within species. It raised interesting questions in particular as to how individual learning and social setting variables peculiar to a group might function in determining the outcome of selection for particular strategies of behaviour. In the next chapter we will discuss the differences and the connections between those blind strategies and tactical decisions. To students of primate behaviour, perhaps especially to those interested also in the evolution of characteristically human behaviour, it is difficult to imagine a sociobiology or behavioural ecology – that is, a modern view of social behaviour – without primates.

Notes

1 The terms *behavioural ecology* and *sociobiology* have in the past sometimes been used almost interchangeably; however, a definition that clarifies relationships between both of these and yet a third characterization of the current approach in some of its aspects is this: *Behavioural ecology* is that broad set of data bodies, methods, and theoretical models of which *sociobiology* and *socioecology* comprise major subsets, and around which evolutionary biology is undergoing reorganization. *Sociobiology* is directed primarily towards explaining the ultimate origins of behaviour in evolution, and *socioecology* toward understanding the operations and dynamic interactions of proximate constraints (cf. Crook 1989). Both take as their starting point the assumption that individuals compete with one another for local resources, with success measurable in terms of relative increase in reproductive fitness.

2 It is important to distinguish between *behavioural recognition* and *conceptual categorization* of kin. What matters from the standpoint of kin recognition theory is that animals act as though they recognize kin (cf. Bekoff 1981), not whether they recognize kin in a way that would allow them to reflect on the nature of kinship. The transition to conceptually informed recognition of kin in human societies will be of interest to us in later chapters.

4 *Emphasizing individual benefits: tactical decisions*

1. Introduction

How do the strategies of the gene get translated into behaviour by individuals and realized in social action? The degree to which responses are invariant or relatively unaffected by contextual constraints differs by species, but all animal species are characterized by a remarkable capacity for adjustment of at least some behaviour to differences in proximate conditions. Evolutionary biology and moral philosophy alike (to say nothing of common sense) require that we view the behaviour of primates and other animals as governed in most instances by conditions which take into account the complexity and diversity of surrounding circumstance. But can the strategies of individuals, as opposed to genes, really be analysed in terms of some definable set of principles on which tactical decisions are based? If the answer is Yes, then how many such principles are there, and what are they like? (We leave the rather different question of how decisions get made until the next chapter).

Anyone who has opened a book on chess in hope of improving his game or who has tried to discover in histories of warfare what principles of command underlie a general's decisions will recall, first, how simple and few are the strategies associated with those contests and, second, how difficult it appears to be nevertheless, even for grand masters and for generals in the field, to bear clearly in mind the few simple strategies which *have* been put forward, to keep them sorted for timely use, and to adhere to them in the hurly-burly of action. Individual life, viewed as an episode in the evolutionary game of survival and reproduction, is certainly not less complex or less mysterious than either of these short-term, contextually limited encounters. To what extent can we hope to gain insight into the social lives of primates, and apply those insights to understanding our own lives in particular, by examining the supposed rules and principles, strategies and tactics, of primate behaviour, not in the abstract but in action. In this chapter we shall consider some of the difficulties such an examination presents, beginning with problems of definition, and then take a look at the kinds of social behaviour to which theories of behavioural strategies in evolution and rules for individual decision-making in real time have been applied.

2. General rules and specific actions

We can begin by asking whether genetic 'strategies', ecological 'principles', and general 'rules' of behaviour can in fact be defined in ways which will afford exact prediction of outcomes in evolution or in day-by-day behaviour. Also, how useful is the metaphor by which we bring evolution into comparison with strategic games? Individual lives are short and, as far as we know, non-recurrent. Questions of punctuationalism versus gradualism aside, evolution proceeds at a slow pace and by incremental change from one generation to the next. This makes it difficult to imagine individuals in the role of active players bent on winning a game called Evolution. Nor, given the emphasis which current theory places on selection at the individual level, is it easier to imagine groups, demes, larger populations, or species in that role. We may observe, following Darwin (and without assuming that life consists solely of competition for resources or that it can be explained simply in terms of such competition), that the struggle for existence is continual and competitive at every level of organization – individual versus individual, group versus group, deme versus deme, etc.; we are not likely to think even metaphorically of evolution going on exclusively at just one of those levels. Nowadays, however, the game analogy is often applied, and seen as having its most useful application, at the level of the gene (Dawkins 1976, 1980; Maynard Smith 1974, 1976; Parker 1984.) We have discovered the virtues of a metaphorical model in which individual organisms are regarded strictly as vehicles of genetic cooperation, 'survival machines' in which 'colonies of genes' compete for resources with other, similar colonies (Dawkins 1976, 1982). In this gene's-eye view of behavior, competition is 'fierce', the struggle is 'relentless', and selection at some point puts 'a premium on central coordination rather than anarchy within the communal body' (Dawkins, 1976:50). Originally, in this view, survival machines provided little more than protective walls for the communal bodies that occupied them, but over time they 'evolved more and more ingenious tricks to increase their efficiency in their various ways of life', the most ingenious of which surely was the acquisition of collective individuality, moving, acting, *thinking* as a coordinated, independent whole (Dawkins 1976:50–51).

Dawkins' portrayal of the social evolution of the gene is vivid and anthropomorphic. The anthropomorphic qualities assigned to genes are central to the metaphor. Genes of course do not sit around in council devising strategies for the future, nor would anyone seriously propose that they do. It is useful, however, where we cannot describe in detail the ultimate origins and evolutionary background of behaviour in which we are interested, and which we are determined to analyse systematically, to construct metaphorical models. In this case the behaviours of concern appear to involve something like cost/benefit analysis and rational selection of alternative courses of action, so it is useful to make the metaphorical model anthropomorphic. Genes are immortal in a sense, persisting generation after generation, each copy the same as the last, and so for a metaphorical game called the Evolution of Social Behaviour they make a good strategic player. One might say despite the fact that they lack a brain, but really it is their constant brainlessness that makes them (collectively, and viewed in the abstract) a good player and allows us to accept the anthropomorphic qualities attributed to them. They play by the rules and not by whim. Happening on a strategy that proves success-ful, they will – we can count on them – continue to employ it until external circumstances require a change; they won't forget it, tire of it, or paint it to suit a mood.

In short, they will behave just as we wish them to, in ways consistent with the model devised for them. But how can we attribute to genes metaphorical motives and thought processes resembling our own, recog-nizing all the while that, in actuality, the character of genic behaviour and (perhaps especially) of interactions between genes is entirely different from that of individual organisms? In fact, it is *because* the behaviour of genes is so different in character from that of human beings and does not embody or reflect directly and immediately the behavioural decisions of individual organisms that we can anthropomorphize genic action without confusing ourselves, at least not to the extent that we do when we attribute human motives and human thought processes indiscriminately to the non-human primates and to other animals, or when we treat as individuals collocations or organized groups of individuals, each with his or her own brain.

So the notion of an evolutionary 'strategy' pursued at the level of the gene is useful for constructing very general models of behaviour in evolution, e.g., models of kin and reciprocal altruism, optimal foraging, etc. Such models incorporate or suggest similarly general rules for individual behaviour consonant with the over-riding goal of maintaining reproductive fitness. It is not difficult to think of what these rules should be like. One can easily draw up a set and give them formal appellations so that they read like *Principles of Individual Action*, e.g.:

1. Keep to your own kind. (Principle of conspecific discrimination.)
2. Don't make friends with (mate with) just anyone. (Principle of social (mating) selectivity.)
3. Stick to the old ways. (Principle of conservative action.)

But . . .

4. If what worked yesterday doesn't work today, try something else. (Principle of generalization and substitution.)

Or . . .

5. If nothing seems to work, try something entirely new. (Principle of innovation.)

Above all . . .

6. Do *something*, even if it's *wrong*. (Don't-just-sit-there principle, or Rash's Law. We of course do not claim to have originated any of these exemplary principles, and though Rash's law was drilled into one of us (Quiatt) by an early teacher, Clifton Rash, we do not suppose that it was originated by him.)

And, finally . . .

7. Where any action appears to invite disaster, do nothing. (Wait-it-out principle, alternative to number 6 and good for the short run only.)

Such very general rules, framed as guidelines for thinking strategists, have not been of much help so far in modelling behaviour, in part it would seem because, lacking a unified theory of cognition, there can be no agreement as to the level of generality at which they should be formulated. As useful in life, and perhaps more closely representative of how the mind functions in learning from experience and in reviewing information and arriving at tactical decisions, are *condition/action rules* specific to particular situations and problems with which an individual is familiar: e.g., if, with respect to the situation at hand, conditions include A and B, but not C, do behaviour x. Rules 5 and 6 above might be framed as decision/action rules in which conditions are set partly by the problem at hand, partly by the outcomes of previous trials. The specific character of any such set of rules ingrained in life clearly must depend, for a given individual, on his or her particular history. At the same time, commonalities across individual sets must be numerous within a species and must increase with the increase of experience common to individuals who have shared the same habitat and perhaps the same group social context.

These implications for learning aside, condition/action rules have played an important role in the modelling of artificial intelligence (AI), and it has been suggested that an inventory of five to six thousand should be sufficient for coping with the ordinary problems of getting around and maintaining oneself which the 'average' human being is likely to encounter in the course of a day (Pylyshyn 1984). One can argue either that these are not so many rules, everything considered, or that they are a great many. The number of rules for behaviour must be great where the object is to connect a specific set of proximate conditions with a single most appropriate action for each. A major issue in artificial intelligence research has been whether the search for *general* principles of cognition should be concentrated on devising routines for locating and implementing appropriate condition/action rules out of some hundreds or thousands of 'correct' rules wired into a knowledge-based 'expert system' program or on building into more flexible programs some combination of problem-solving and learning routines which could enable a simulated intelligence to structure its own memory and in some sense fashion its own rules (Haugeland 1985). Another computational approach to simulation research in cognition, which in some respects, and certainly by self-representation, is 'anti-AI', is neural networking, in which general principles are, in effect and indeed almost literally, wired directly into the circuitry of connectionist models (for reviews of the controversy and descriptions of neural networking see contributions to Graubard 1988.) Here our primary concern is to note, with respect to expert knowledge systems, that wiring into a system a great number of different sets of nested conditions, along with the background information necessary to effect transitions from state to state through each set toward appropriate (end state) resolutions in action, requires immense stores of memory which are unlikely to be used efficiently. Where condition/action rules are oriented toward highly specific applications, as in expert system programs, a great number must be needed for full description of even a relatively simple society or social subsystem. An explanatory model of behaviour, it would seem, must be general if it is to be useful or even interesting; but, being general, it is unlikely to afford reliable prediction of outcomes of specific behavioural events. General explanations nevertheless are to be preferred over highly specific explanations of behaviour (Dawkins 1989).

Unfortunately, behavioural scientists often do give the impression of anticipating that genetic logic models will enable prediction of behaviour, not directly perhaps but at least when allowances are made or corrections calculated for translation of gene action into whole organism behaviour (e.g., Lumsden & Wilson 1981). In some few instances, organismic behaviour *can* be analysed in terms of genetic input, but, in general, the regularities to be observed in complex social behaviour appear to be influenced by a great number of variables, external as well as internal to the individuals involved. Attempts to explain social behaviour by categorizing systematically alternative outcomes on the basis of variable genetic input seem to be doomed from the start if this is not taken into account. As Pattee has observed (1970), 'it is easy to understand how a simple change in a single variable can result in very complicated changes in a large system of particles. This is the normal physical situation. It is not easy to explain how complicated changes in a large system of particles can repeatedly result in a simple change in a single variable.' This, so one supposes, is the normal social behavioural situation.

Strategy and tactics

In ordinary (i.e., not specifically biological) discourse, the terms 'strategy' and 'tactics' are used synonymously or, more often than not, the latter term is used to refer to a particular goal-oriented disposition of resources consonant with an encompassing strategy. This raises expectations about their association in scientific contexts, where in fact they are rarely assigned specific contrastive definitions (Dunbar 1988, and see below). A 'strategy', one may assume, governs operational use of certain 'tactics' out of some larger number available for use; selection of a particular tactic is consonant with underlying strategy. One of the seldom mentioned problems which arises almost inevitably when genetic logic models are applied to the analysis of behaviour in evolution is that it is very difficult to talk about the application of the model to observable behaviour without appearing to have mixed levels of analysis. At one level of analysis it is customary, after noting the apparent functions of a behaviour, and relating these to a likely past history of selection and genetic adaptation, to refer to the *ultimate origins* of that behaviour; at another level it is useful to identify the *proximate conditions* responsible for manifestation of the behaviour. Few would argue the utility of trying to link, by way of explanation, ultimate origins of some anatomical feature directly and explicitly with its proximate manifestation in a given individual, but confusion of levels of analysis is likely to occur where the tactical decisions of behaving organisms are viewed as governed directly or indirectly by named strategies of the genes. It is necessary then to bear in mind that the selfish gene model of behaviour in evolution is metaphorical as well as anthropomorphic. The main problem with it lies in the increasing difficulty we are likely to experience in distinguishing between metaphor and reality as selfish gene explanations are spun out (see Crook 1989, for discussion of this and other weaknesses of sociobiological method.)

Of particular interest, where the evolution of social behaviour is concerned, is the extent to which real decisions concerning alternative forms of behaviour may be influenced by cognitive review of relevant information and consideration of tactical questions. From the perspective of scientists interested in human evolution and in the interplay between cultural and biological selection (e.g., Plotkin & Odling-Smee 1981; Boyd & Richerson 1985; Rindos 1986), it is important to ask how decision-making of this sort may vary across species, and what the rules or principles are which govern the making of such decisions. This is quite a bit different from asking what the rules or principles are which govern behaviour which is not identifiable as involving intentional action – or in relation to which intention cannot reasonably be assumed. At this point we are more interested in considering behaviour *per se* than in trying to distinguish the two kinds, hence we postpone to a later chapter definitional discussion beyond these few remarks. For the present, we follow Dunbar (1988) in rejecting the convention that defines strategy as 'genotypic responses made by members of a particular species as a whole', and tactics as 'the essentially phenotypic options open to individuals'. Dunbar prefers the convention of everyday discourse

> which views tactics as a set of choices specific to a particular strategic option. This leaves us free to use the terms in a strictly relative sense so that what at one level of analysis is a tactic within a particular strategy with its own array of tactical options at a lower level of analysis'. (Dunbar 1988).

This solves the problem of mixing levels of analysis, and it seems wise to give up that beguiling shorthand by which an underlying 'genotypic' strategy of the species gives rise to the environmentally constrained and phenotypically variable tactics of individuals. The ultimate task of behaviour analysis must be to see how genetic logic and ecological principles influence individual cognition and decision-making, but conflation is not the answer to the question of how they fit together.

What is perhaps most important to keep in mind is that the rules, principles, strategies and tactics of individual decision-making are not the same rules, principles, strategies and tactics which we are likely to have in mind when discussing genetic logic models of social behaviour in evolution. At both levels, but at the level of individual action most obviously, 'The rules that govern behaviour are not like laws enforced by an authority or decisions made by a commander: behaviour is regular without being regulated' (Gibson, 1979:225). 'The question,' says Gibson, 'is how this can be'. We are not ready to tackle that one, but a prior question is simply what *are* the regularities of behaviour, and as field primatologists we are interested in that question, in particular as it concerns social behaviour. It is worthy of interest not just for its own sake but because if there are no laws to inform us what the outcomes of behaviour must be, we can only establish what those laws might be like, if they existed, from consideration of outcomes observed (Kummer *et al.* 1990).

3. The organization of social behaviour

As we noted in Chapter 3, the questions that ethologists ask about behaviour are of several kinds: How does the ability to behave in a certain way develop? What are its immediate causes? What is its function? How did it evolve? In studies of primate social behavior the main emphasis in recent years has been on function, that is, on evolutionary function defined in terms of selection and adaptation and viewed from the standpoint of individual animals competing with one another for mates and resources. This emphasis on individual competition and individual selection does not mean that we can achieve comprehensive understanding of group social behaviour by seeing it as simply the sum total of individual behaviour, or that we may usefully identify the group-as-social-*system* as, at best, a kind of self-integrating or free market system of manipulative individuals bent on maximizing reproductive gain. However, for purposes of initial description, record, and summarizing analyses, a model of social behaviour that is primarily organizational and simplified along these lines is both conventional and useful.

A particularly well developed theory of the organization of social behaviour, beginning with the behaviour of individuals in dyadic interaction, has been outlined repeatedly by Hinde (e.g., 1976, 1982, 1987). Hinde's theory, though rooted in traditional ethology, is descriptive not so much of the forms of behaviour (e.g., of communicative displays) – those appear to be taken for granted – as of its overall structure. Understanding of a social system, says Hinde, is contingent on an adequate descriptive base, and for that 'an absolute requirement is clearheadedness about the levels of complexity involved' (Hinde, 1982). Hinde focuses on four levels which he sees as distinct in terms of analytic components and procedures: individual behaviour *per se*, interactions between individuals, individual relationships, and overall social structure (See Figure 4.1). Analysis at each level must deal with 'properties that are not relevant to the level below'. Perhaps because Hinde's main concern is with the description of how behaviour is organized, not with ultimate origins, he does not attend to levels that underlie whole organism behaviour, beginning with gene action.

Nevertheless, Hinde's scheme, with its emphasis on dyadic interaction and relationships arising therefrom, is nicely consonant with sociobiological models of competition and cooperation (Clutton-Brock & Harvey 1976). Hinde's theory has been influential in part because of his attention to the hierarchical organization of social behaviour. That influence is evidenced in derived conceptual models (see, e.g., Colmenares & Rivero 1986) as well as in methodological applications (by, not least, Hinde's many students; see the numerous contributions to Hinde 1983c). It is easy to see why a scheme such as Hinde's should appeal to field primatologists engaged in compiling detailed accounts of who did what to whom, when, and where. Field primatologists are habituated to asking how patterns of social interaction over time differ from individual to individual (see Chapter 3), and to looking for clues to larger group structure in the links between (on the one hand) spatial association and interactional patterns and (on the other hand) individual age/sex class and genealogical or other sub-grouping. In short, Hinde's four levels reflect the procedural stages of defining observational units of behaviour, recording interactions, summarizing data over sessions of record, and arriving at generalizations concerning group social organization.

4. The character of social relationships

We do not mean to suggest that Hinde's scheme has been influential only because it suits established practices of fieldworkers in primate social behaviour. More important is the emphasis which his scheme places on *relationships* as the central elements, the connecting members of social structure. A record of the frequencies with which interactions are initiated by either party (and terminated, or avoided) is of course basic to description of a relationship. Differences in duration and intensity of particular behaviours are important for more subtle comparisons of quality, e.g., in relationships between mothers and their offspring in a group of baboons (Chalmers 1979). Measures of duration and intensity are useful too for comparing different kinds of relationships, for instance between siblings of like sex and siblings of unlike sex, siblings whose ordinal positions are contiguous and siblings between whose births another birth or other births have intervened, etc. (e.g., Datta 1989). Ideally, one would like to record behaviour in all its dimensions, but to employ more measures than those of frequency and rate, duration, and

88   *Emphasizing individual benefits: tactical decisions*

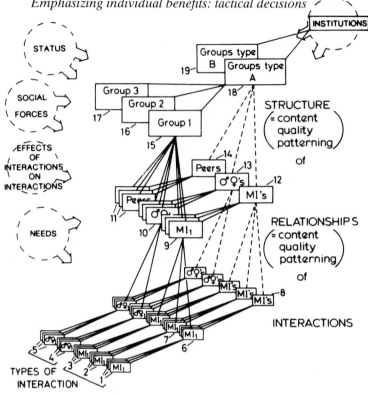

Figure 4.1.   Interactions, relationships and social structure are shown as rectangles on
three levels, with successive stages of abstraction from left to right. The discontinuous
circles represent some of the principles necessary to explain patterning at each level.
Institutions, having a dual role, are shown in both a rectangle and a circle. 1. Instances of
grooming interactions between a mother A and infant B. 2. Instances of nursing interactions
between A and B. 3. Instances of play between A and B. 4. Instances of grooming between
female A and male C. 5. Instances of copulation between A and C. 6. First stage abstraction
– schematic grooming interactions between A and B. Abstractions of grooming interactions
between other mother–infant pairs are shown behind, but the specific instances from which
they were abstracted are not shown. 7. First stage abstraction – schematic nursing
interactions between A and B. Abstractions of nursing interactions of other mother–infant
pairs are shown behind. 8. Second stage abstraction – schematic grooming interactions
between all mother–infant pairs in troop. 9. Mother–infant relationship between A and B.
Mother–infant relationships of other mother–infant pairs are shown behind (but connections to grooming, nursing, etc., interactions are not shown). 10. Consort relationship
between A and C. Other consort relationships are shown behind. 11. Specific relationships
of another type (e.g., peer–peer). 12–14. Abstraction of mother–infant, consort and peer–
peer relationships. These may depend on abstractions of the contributing interactions.
15. Surface structure of troop containing A, B, C, etc. 16, 17. Surface structures of other
troops (contributing relationships are not shown). 18. Abstraction of structure of troops
including that containing A, B, C, etc. This may depend on abstractions of mother–infant,
etc., relationships. 19. Abstraction of structure of a different set of troops (from another
environment, species, etc.). Rectangles labelled MI$_1$ refer to behaviour of dyad female A
and her infant B. Rectangles labelled ♂♀$_1$ refer to consort pair female A and male C.
Rectangles labelled MIs, ♂♀'s refer to generalizations about behaviour of mother–infant
dyads and consort pairs respectively. (From Hinde 1982, by permission of Oxford
University Press.)

intensity would soon have the recorder floundering in detail, especially in the case of focal group recording sessions in which time is of the essence. Nevertheless, difficult though it may be to describe interactions except by rather basic categories and frequencies, and dangerous too to dwell too lovingly on differences in individual style, multi-dimensional characterization of behaviour is an appropriate goal where circumstances permit and where it can be managed with something like objective rigour.

Analysing social relations

Hinde (1987) makes a useful distinction between the surface and the underlying structure of a social system. Discussion of the relationship between any particular mother N and daughter n entails description of the mother–daughter relationship at the level of surface structure. Abstracting elements common to interactions observed within dyads of this type affords description at the level of the underlying structure of properties inherent in the relationship between mothers and daughters in a given group (or deme or species, depending on the aim of abstraction and summary). That reference to 'abstracting elements' reminds us of what too often is conveniently ignored, that analysis can only deal with observational units, not with the transient behaviour to which they have been applied. With this in mind we can note that the distinction between surface and underlying structure makes it possible to treat categorical relationships as something like the connecting members of an observer-built structure of social organization. We put it this way, however awkwardly, to emphasize the descriptive character of the theory.

Hinde's distinction between *surface* and *underlying* structure may seem obvious, just a way of moving from summaries of interactions between particular individuals to something like generalized role statements, but it has important uses. Primarily, it affords systematic, intensive comparison of relationships within and across categories. Sequential recording of interactions and close description of the surface structure of a relationship 'fixes' dyadic behaviour in its temporal dimension, and long-term compilations of bouts and sequences of interactions provide the units for analysis of underlying structure (an appropriate analogue is reels or tapes on which are captured long series of interactions between the same two individuals). Attention to duration and intensity of communicative signals, and to qualitative analysis in general of surface features of a relationship, are of particular importance in the comparative assessment of underlying structure. Thus there is a direct analytic connection between Hinde's surface and underlying structures; there is, here, nothing like the conceptual gap between observable and (hypothetical) underlying constructs which economists, linguists, and social anthropologists occasionally bridge by reference to the 'deep' structure of behaviour. Also, relationships sum up activities of particular importance, though it probably would be better to say that primatologists attend to and designate relationships that concern behaviour in which they have an *a priori* interest and which they perceive as peculiarly important to structure – e.g., within-sex competition, mating activities, mother–offspring interactions. In any event, in the study of relationships attention is directed to what are generally perceived as important functional links between individual interactions and the structure of the group as a whole. Hinde's descriptive approach, linking the behaviour of individual organisms with dyadic structures, and those structures with larger sub- structures of the group, in effect skirts the issue of individual versus group selection. The description is nicely consonant with individual selectionist theory, beginning as it does with the behaviour of individuals, but it may turn out that behaviour of the larger group hinges sometimes on dynamics that are not revealed by description that follows this route.

What determines the character of a relationship? This question is most easily answered at the level of surface structure, where an observed history of interactions between particular individuals can reveal the workings of identifiable proximate constraints. At the level of underlying structure, though there are no easy answers to questions concerning cause, we at least are in a position to analyse correlations between differences in role-related behaviour and genetic variability in groups of known genealogy. Thus a distinction between surface and underlying structure seems to facilitate directing attention usefully toward proximate conditions of behaviour on the one hand and ultimate limits of genotype on the other.

Because they exist in time and are influenced by past interactions as well as expectations of the future, relationships are unlikely to remain static but will vary within limits, 'chang[ing] . . . progressively with time, or from one temporarily stable state to another' (Hinde & Stevenson-Hinde 1976). This variability in conjunction with the retention of a certain constancy of character over time is of particular concern to Hinde. If relationships are not static, neither is the structure of a group as a whole, and the study of social structure essentially involves a search for the principles which explain how behaviour is organized in relationships – more precisely, how relationships are maintained in dynamic stability over time (cf. Humphrey 1976). For Hinde, principles which apply to the dynamic stability of relationships are of several kinds, some having to do with propensities toward relationships within or between classes (e.g., age, sex, status, kinship), others with the role of learning in the formation of relationships, still others with the character of interactions *per se*, character of interaction sequences, and degree of compatibility between actors in a relationship. Given the dynamic character of interactions and relationships and the multiplicity of properties both that are internal and (from a broader ecological and relational standpoint) external, it is hardly surprising that Hinde nowhere puts forward a definitive list of explanatory principles. He has suggested however that:

> In practice the principles necessary for understanding the patterning of interactions within relationships can be classified into three interrelated groups:
> (a) Principles concerned with the natures or needs of individuals . . .
> (b) Principles concerned with the effects of interactions on interactions . . .
> (c) Principles concerned with social influences on the dyad . . . (Hinde, 1982, pp. 209–10)

It is not clear whether Hinde attaches any special theoretical import to this classification. Although its organizational properties are apparent, it seems to function primarily as a device for convenient reference, a tentative listing of principles which Hinde regards as being particularly useful. In terms of the analogy we introduced just above, those videotapes of structure seem intended primarily for storage and retrieval of information – but note how they might figure in an informational theory of culture, where the storage is conceived of as neural on the one hand, and social (and increasingly, in human evolution, technological) on the other, and the retrieval is that of individual cognition or institutional activity.

Hinde has suggested elsewhere (1976) that the search for an adequate general description of social organization and the search for explanatory principles may meet in the study of human institutions, and he has devoted considerable attention to the study of human behaviour, in which 'the institutionalization of relationships introduces a new dimension of complexity' (Hinde 1976). The institutionalization of relationships should have important consequences for the emergence of human culture, including value systems in dynamic relationship with complexly integrated sets of social expectations linked to conventional but non-arbitrary rewards and sanctions. However, in the work that represents his most thorough attempt to unify ethological and sociological explanations of behaviour from a 'relationships' perspective (Hinde 1987), Hinde does not elucidate clearly the processes responsible for that institutionalization.

We agree that the institutionalization of relationships was critical to human evolution (see Chapters 9 and 10). The problem is how to define that institutionalization in a way which will open it up to the kind of close analysis that Hinde has accorded relationships *per se*. A definition which is simple and operational (or at least consistent with a perspective that lends itself to operational procedures) is best for this purpose, it seems to us. A relationship serves participants as a two-way repository of information. It also serves informational functions not only when human observers employ it as an analytic concept but also when animals observed use it in a similar way – that is, as a cognitive construct connecting the neural system with patterns in the external world which are congruent in some but not all respects for conspecific beholders of an interaction as well as for those engaged in it (Cheney & Seyfarth 1980; Dasser 1988; Seyfarth & Cheney 1988). Primate social behaviour, to the extent that it involves recognition and manipulation (again, conscious or no) of relationships in interaction with others, must require processing of quite complex bodies of information.

5. Culture as an informational construct

We propose to define the *institutionalization of relationships*, as:

1. *conventionalization of form*, facilitating recognition of a wider set of relationships, and
2. *conventionalization of processes* by which a relationship contributes to the ability of individuals to store, retrieve, and in general make use of social or other intelligence.

However, we leave discussion of how relationships *may* have become institutionalized in human evolution for a later chapter. We must first develop the concept of culture as information, a concept which would seem on the face of it not to be tied restrictively to consideration of just one species, our own. In our view, an information-based theory of knowledge should be applicable to all forms of life (see Staddon 1989: 'Psychology should be the study of *intelligence* of adaptive and complex behaviour, wherever it is to be found').

This view runs counter to current biases associated with two areas of theory which are important to understanding the evolution of human culture in its relation to primate social cognition. One of these is the anthropocentrism which has so far characterized neural simulation and computational approaches to cognition, in which models of 'artificial intelligence' are almost without exception presented as analogues of *human* intelligence. This bias can probably be explained in terms of the history of research in simulated cognition, which has always had one foot in the camp of philosophers, linguists, and computer scientists concerned with symbolic logic and problems of language translation, and the other in a camp of electrical engineers, neurologists and psychologists concerned with pattern perception, neural circuits, and behavioural control systems – with competition keen within as well as across camps. In both camps the main task has been to construct machines which can replicate or at least inform us about human action. To outsiders like ourselves contributions to date seem for the most part as applicable to the cognition of mammals in general as to intelligence that is specifically human. Hence, we deal with the bias toward human cognition by ignoring it.

The other bias, which cannot be so simply written off, reflects an anthropocentric view mainly in the odd, backhanded way in which it posits a crucial gulf between animal and human nature. It involves denying to animal communication the informational character which we tend to regard as central to human language (Dawkins & Krebs 1978; but cf. Griffin 1976, 1984; Lyons 1988). If this view is correct, that information about the surrounding world is just what is missing from animal signals, then clearly there are difficulties in store for anyone who proposes to apply an informational definition (of the nature of an individual relationship) to the comparison of human behaviour with that of other primates. So a first task is to see whether (or for what purposes) we may be justified in rejecting proposals like that of Dawkins & Krebs, that in analyses of non-human animal communication signals be regarded as having no informational content. In the next section we examine their argument and give our reasons for preferring an informational model of animal communication.

6. Communications and the management of relations

Primate social behaviour, in evolution as in everyday occurrence, in our view is responsible for but also is contingent on transmission of information – via the genes of course, in evolution, and in daily life through interactive communication (in conjunction with endogenous, gene-linked processes) between individuals who maintain or wish to initiate or terminate relationships with one another. This may seem self-evident: the idea of particulate inheritance, which in essence is a system of exchange and dissemination of genetic information, is basic to our understanding of evolutionary biology, and the study of social behaviour is in many respects equivalent to the study of communication between individual organisms. Discussion of the organization and structure of social behaviour, in human beings and other animals, must begin with the consideration of current disagreement concerning the functions of communication. A basic model of communication may be diagrammed as in Figure 4.2, with transmitter and receiver, a transmitted signal, and feedback. Analysis of communication, even in a simple one-dyad system, must usually take into account features such as the organs of signal transmission, reception, amplification, and conversion for storage, the medium over which a signal is transmitted, and environmental factors which distort the signal or introduce noise into the system. Of central importance to communications theory are feedback mechanisms that enable the sender of a signal to monitor responses to its reception and alter sending behaviour appropriately. Analysis of feedback affords human observers operational measures of the meaning of specific signals, a fact which is not incidental to arguments about the nature of animal communication.

The same communicative event can be analysed from the standpoint of a sender interested in affecting the behaviour of a receiver or from that of a receiver tuning in on the news, that is, on information about changes in the external world which may have a bearing on her or his well-being – and we note that changes in the environment external to the receiver include changes in the state of mind of the signal sender (cf. Dawkins & Krebs 1978.) In any case, analysis of a communicative event from the standpoint of individual organisms involved, and in terms of the evolution of social behaviour, must eventually take into account what sensory data are utilized, how cognitive integration proceeds, what rules may govern evaluation of costs and benefits, and what constraints are likely to be imposed on action.

Ethological models of communication have tended to focus on the *sender* of signals. Although ethologists may appear to define communication even-handedly as including 'any stimulus arising from one animal and eliciting a response in another' (Scott 1968), attention to responses, and to the character and functional implications thereof, usually turns out to be primarily a way of asking whether a signal has in fact been sent and how the context of action is thereby changed for the sender. In many instances, of course, the functional meaning of a signal can *only* be inferred from the response that it elicits (Tinbergen 1951). It is ironic that our own limitations as observers, which require us to define the meaning of signals in terms of observable responses elicited, should bias us against attributing volition to responding organisms. Selfish-gene theory provides a theoretical rationale for assigning priority to the signaller, and the logical consequences of so assigning priority have been developed by

(a)

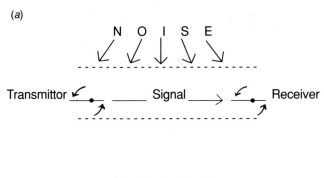

(b)

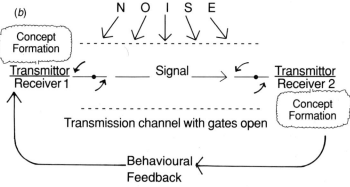

Figure 4.2. Communication system in outline. (a) Signal transmission without feedback. (b) Signal transmission with feedback in a two-way communication system characterized by cognitive transformation. In this system reformation of TR1's initial concept or repatterning of signal or both may follow observation of the behaviour of TR2, which is assumed to depend in part on reception and interpretation of the original signal.

Dawkins and Krebs into a rigorous definition of communication as manipulative behaviour. Communication so defined is, they claim, 'so far from the spirit of what is conveyed by the ordinary usage of the word that we are tempted to abandon the word communication altogether' (Dawkins & Krebs, 1978). But of course they do not abandon it, and that is where the problem lies.

In the view of Dawkins and Krebs communication benefits the *actor*, that is, the signal sender, and any benefit to the receiver of the signal, in their terms the *reactor*, is coincidental. Signal transmission is a substitute for force in exploiting the physiology and anatomy of another individual, even the 'sense organs and behaviour machinery which are themselves designed to preserve the genes of that other animal' (Dawkins & Krebs 1978). Better said, perhaps, it applies in a highly efficient manner a particular *kind* of force to such exploitation: Dawkins and Krebs cite J. B. S. Haldane's remark that 'a general property of communication is the pronounced energetic efficiency of signalling: a small effort put into the signal typically elicits an energetically greater response' in a manner 'reminiscent of electronic amplification' (the citation is from E. O. Wilson 1975). Dawkins and Krebs offer their manipulative model of communication as an alternative to a particular information model, that of 'classical ethology' (see e.g. Smith 1977 and, for Smith's response to Dawkins and Krebs, Smith 1984). However, as we have noted, the analytic focus is on the signal sender in both Dawkins and Krebs' sociobiological model of communication and Smith's 'classical ethological' model; the difference between them lies in whether or not signals are viewed as having informational content.

Dawkins and Krebs explicitly reject 'informational' interpretations of communication; this rejection appears to be the reason for their choice of terms: *actor* and *reactor* instead of, e.g., *sender* (or *transmittor*) and *receiver* of signals. They have not been concerned, by and large, with the pattern content of signals or with the ideas that a receiver may entertain about that content – the qualifying phrase is necessary because the 'mind reading' that they predicate in the second of their two co-authored papers suggests something at least formally equivalent to an informational review by receivers of signals received. Nevertheless, where signal content is concerned, Dawkins and Krebs 'eschew . . . ideas of information and of meaning' (1978), preferring in most (not quite all) cases to think of the communicating animal as 'manipulating' the behaviour of its companions. Their view of animal communication, manipulative and non-cognitive, is elegantly conceived and powerful, an efficient tool for many purposes of behavioural analysis; it is perhaps least useful for understanding the evolution of primate cognition and of human communication.

In the view of life as a contest between selfish genes, a compatible model of communication at the individual level naturally emphasizes attempts by the displaying individual, the signal sender or actor, to manipulate others toward the achievement of his or her own maximum reproductive fitness. In this view, communication benefits the actor, and complex issues of cognition and self-consciousness relating to the signal as message – whole areas of meaning connected with selective reception and interpretation by the signal receiver – are intentionally left out of the account. Such issues, which may be central to the concerns of anthropologists and other students of human behaviour, did not until very recently figure importantly in the research into non-human primate social behaviour. Anthropologists have tended to find transactional models of social interaction (e.g., Homans 1958, 1961) or even dramaturgical models (Goffman 1959, 1974; Harré 1979) more useful as bases for their research than a strictly manipulative model. In a transactional model of interaction, transmission of communicative signals with feedback can be likened in its reciprocal aspects to market transactions in which exchange of goods and services work to the benefit of all parties. The transactional programme of exchange theorists in social anthropology has been adapted to analysis of nonhuman primate communicative behaviour by Chadwick Jones (1987a,b, 1989a,b) in a way that seems to us most productive (Figure 4.3). From a transactional standpoint, conspecific communicative events (informational content of signals aside) are understood to be, by and large, acts of reciprocal exchange, and elimination of reciprocity may be regarded as a failure in communication.

The analytic focus of informational models of communication has traditionally been on signal content for various reasons. One has to do with the fact that development of information theory in the 1940s had an important impetus in wartime and postwar research (much of it in the Bell Telephone laboratories) centred specifically on isolating informational properties of electrically transmitted signals (Shannon 1949); another has to do with longstanding interdisciplinary concern with bee 'language' and with communication features of 'messages' communicated thereby (von Frisch 1954, 1967; Lindauer 1961); still another with the way elements of information theory have been used in classical ethological explanations of the ritualization of behaviour in evolution (Tinbergen 1951; Smith 1977). With regard to the last, movements (display elements) incidental to emotionally motivated behaviour are conceived to have been shaped stereotypically (formalized) via natural selection so as:

> (a) to promote better and more unambiguous signal function . . . ;
> (b) to serve as more efficient stimulators or releasers of more efficient patterns of action in other individuals; (c) to reduce intra-specific damage; and (d) to serve as sexual or social bonding mechanisms (J. Huxley, 1966, cited by Dawkins & Krebs, 1978).

The emphasis here on reduction of ambiguity 'makes sense only in the context of an exchange of information' (Dawkins & Krebs 1978) and is not really compatible with an agonistic view of life and a manipulative

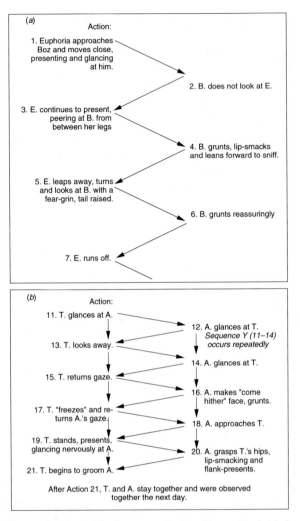

Figure 4.3. Transactional model of communication (From Chadwick-Jones 1989b). (a) Reactive contingency, (b) mutual contingency. Compare (*a*), in which responses are tentative and perhaps superficial, with (*b*), in which a transactional relationship is pursued.

behaviour model of communication. The notion of information exchange, Dawkins and Krebs suggest, is all very well for human language, 'where the end result is that the receiver learns something which he did not know before, from the sender'. But, they ask, 'in the case of animal signals, what is the "information" supposed to be about?' Applications to bee dances and birdsong *possibly* excepted, 'most "informational" interpretations of animal communication have concentrated on information about the actor's internal state rather than about events in the outside world.'

The Dawkins–Krebs manipulative model of communication involves a limited, consistent focus that is useful when we are interested primarily in asking how, in general, an individual may attempt to manage and integrate relationships toward achievement of selfish goals. Since manipulation of the signal-receiver must involve concealing or disguising aims which work to the disadvantage of the receiver, some important questions are raised about the role of deception in communication (central, it would seem) and about its nature as well (what may deception involve beyond withholding information?). Krebs and Dawkins subsequently (1984) shifted their perspective to argue that while communication can be regarded as strictly manipulative, with selection for increasingly effective social manipulation by the individual-as-actor, there must at the same time be selection for defence against being manipulated to one's disadvantage. For individuals 'aware' of where their own interests lie, understanding of how the deceiving actor's true aims depart from those interests would be sufficient to undeceive them. Hence selection for 'mind-reading', Krebs and Dawkins' suitcase term for whatever features contribute to this perception, should operate in conjunction with and counter to selection for greater manipulative proficiency. The two tendencies, which, they argue, are intimately related and consistently opposed, account for much of the complexity and subtle character of animal social life.

This is undoubtedly an attractive and for many purposes a useful model. There is something very satisfying in the ways it links ideas about the relationship between genes in evolution with ideas about relationships between interacting individuals. However, it is important to keep in mind that it is a selfish-gene model in which individuals are treated simply and solely as vehicles of genetic evolution. It can dispense with information as a useful concept because it was not designed for analysis of problems having to do with cognition in evolution.[1] Also, because Dawkins and Krebs specifically exclude human communication from the domain of behaviour governed by their model, it cannot be employed in cross-specific comparisons designed to shed light on the evolution of language and culture. Selfish-gene models are usefully reductionist; it is by reducing behaviour to terms of genetic logic that they are able to handle with elegance and rigour problems which can be treated at this level of abstraction. To point out that they achieve elegance and rigour by not attending to systematic organization of behaviour at the level of the individual is not a serious criticism of such models.

Nevertheless, the view of communication expressed in these two influential papers (Dawkins & Krebs 1978; Krebs & Dawkins 1984) is a very narrow one, and the view of individual cognition is narrower still. For anthropologists whose interests in animal communication are closely linked with a concern for understanding the evolution of human language and culture, a theory of communication in which selection is viewed as working simultaneously to produce more effective manipulation of conspecific companions by signal transmittors and more effective defence against such manipulation by receivers is unlikely to prove of more than passing interest. The situation is only superficially analogous to those in which coevolutionary selection appears to maintain dynamic equilibrium of behaviour *across* species (e.g., in adaptations for rapid locomotion by predator and prey species or other 'arms races', Parker 1983).

While it may be true that 'most "informational" interpretations of animal communication have concentrated on information about the actor's internal state rather than about events in the outside world' (Dawkins & Krebs, 1978, and see above), the emphasis on differences between 'the actor's internal state' and 'events in the outside world' is interestingly inappropriate to the context in which they are discussed by Dawkins and Krebs. For what is at issue to the receiver is whether he is receiving information that he may use to advance his own interests; in terms of that issue, and from the standpoint of the receiver, the internal state of the signal sender is a real and potentially significant complex of 'events in the outside world'. The significance of this relationship between the signal sender's state of mind and that of the signal receiver, the former representing a system both interesting and useful in the world external to the latter, is bound to be overlooked when we focus our attention too narrowly on the *actor*; in this view the mind-reading *reactor* is at best engaged in passive defence, along a single front, of his own interests (see M.S. Dawkins & Guilford 1991 and Guilford & M.S. Dawkins 1991 for more recent and broader examinations of receiver behaviour and receiver psychology in relation to the evolution of animal communication systems). It is clear that what Dawkins and Krebs are concerned about is not so much information *per se* as *shared* information, or, more precisely, the extent to which a signal may have the same meaning for its sender as for its receiver. This is a practical problem for human users of language and a philosophical problem for detached observers of language in use; it seems to us that the problem is not greatly different in kind when it comes to communication in other animal species. If it can be reduced to a problem of translation, as we would argue it can, then it is only *less* of a problem for our own species, in which the forms of processing socially accessible information have been refined and conventionalized by genetic and cultural adaptation – primarily but not solely in connection with language.

This may be a naively evolutionist view of human language, but it has the virtue of forcing those who adopt it towards a narrow but powerful definition of culture as the product of such processing: *culture = socially processed information*, a definable subset of the environmental (as opposed to genetically encoded) information which is accessible to a given species. We agree with Dawkins and Krebs that analysis of communication as social behaviour is best pursued from a standpoint at one end or the other of the transmission; it is difficult to do it from both ends at once. Where we differ is on whether the question of informational content of signals is the kind of problem they make it out to be. If we suppose that a signal has informational or indeed any kind of content which can be distinguished from the pattern which comprises that signal and which separates it from surrounding noise we are likely to wind up in definitional straits, but surely no sooner than if we suppose that signals emitted by animals are somehow exempt from the kinds of analysis which information theory brings to bear on communication between human beings.

Other animals may be interested in the signalling behaviour of conspecific individuals for reasons quite apart from anything which that behaviour may communicate about a signal sender's aims and strategies, as we are ourselves. Conspecific others, especially those with whom an individual is accustomed to interact and has formed rapport, and increasing mutual sensitivity to changes in physiological state and emotional mood, may serve as something like detached receivers and transmittors, providing continual updates on the availability of resources and the character of environmental circumstances (Quiatt 1984, 1988b; Small 1990). Cheney & Seyfarth's observations (Cheney 1984) of the ontogeny of vervet monkey alarm calls and of receiver's reactions to 'false' alarm calls (Cheney and Seyfarth 1988) provide confirmation from the field of the dynamic character of feedback, i.e., as viewed from the standpoint of the ultimate recipient of information as signal modulator, 'tuning' intermediate conspecific receiver/transmittors for optimal performance. Familiarity breeds understanding; the closer the association the more frequent the interaction and the more use which can be made of such detached sensors. In this respect individuals can be seen as manipulative whether they are behaving as receivers or as senders of signals – a point perhaps implicit in Krebs and Dawkins's notion of 'mind-reading' (1984). It seems unlikely, however, that our understanding of the character of relationships, how they vary, how they are managed over the long term, and how they may be integrated one with another, will be much improved by emphasizing strictly the manipulative aspects of communication. Comprehensive understanding of social relations, at underlying levels as well as at the surface, in functional analyses and especially in an evolutionary perspective, depends on a conjunctive approach which goes beyond the application of selfish-gene models to include consideration of all the ways in which conspecific others contribute to the information base on which an individual's reproductive fitness ultimately depends – an approach which is partly 'sociogenetic' but mostly socioecological.

7. Simulation models of intelligent behaviour

How important the notion of information exchange is to an evolutionary analysis of non-human primate social relationships, in their transactional aspects, may be a matter for debate.[2] For anyone interested in the evolution of culture, including both its preconditions in biological evolution and its continuing transformations, it would seem to be important to begin not with an essentialist view of the nature of language but by assuming a continuum in the evolution of communication from ancestral non-human primates to modern hominids to contemporary and future robotic 'post-hominids'. That would lead naturally to such interesting questions as: 1. whether exchange of information in social communication could have preceded the evolution of vocal language (glottogenesis); 2. whether the evolution of vocal language could have succeeded or even have required (as Gordon Hewes has persuasively argued 1973a,b, 1976, 1977) a prior gestural language; and 3. more generally, how, as the role of informational content in social exchange became more important, the character of that exchange may have been otherwise altered.

Individual cognition may be thought of in its processual aspect as a kind of internal communication (Bickerton 1990); and, whether or not one chooses to think of it in this way, it is difficult to think of it at all without extensive resort to informational terms and concepts. Indeed, the 'cognitive revolution' of the last decade or two seems to be rooted directly in the emergence of information processing as a central concept in communication systems theory. This is true whether one has in mind rapid advances in the study of neural anatomy and physiology of actual systems or in the construction of simulated systems of cognition. In the next chapter we review cognition in actual, real-life primates. Here we will examine some current simulation studies which seem to us of particular interest in considering the evolution of primate and human cognition.

The application of computer technology to the simulation of cognitive processes or, to use the trendier phrase now that guidance systems and industrial robotics are well established, research in artificial intelligence (AI), has so permeated that cognitive revolution to which we have just referred that its major events are chronicled not so much in seminal journal articles and revisions of texts as in the addition of sub-routines to working programs and distribution of new versions of software. A computer program may be 'the most natural way to do what . . . [a] theory [of cognition] itself has to do – namely account for the information processing that goes on in the brain' (Allen Newell 1987, as cited by Waldrop 1988b; see also Newell 1990).

An early computerized representation of the way which information might be processed by human beings applying basic, very general rules of reason to cope with a variety of ordinary problems was the General Problem Solver devised by Newell & Simon (1963). Analysis of human problem-solving behaviour yielded a number of widely used techniques such as trial and error, making decisions on the basis of available knowledge, and, where no immediate solution presented itself, changing the search context in a manner calculated to move one closer to it. This last, means–ends analysis, was one of the cornerstones of Newell and Simons's construction. The General Problem Solver created a problem space with configurational states and operators by which the configuration could be changed from one state to another (Figure 4.4). The solution lay in discovering a sequence of transitions from an entry state to an acceptable end state. Associated with each state was a set of informational elements represented in short term ('working') memory. The operators on which state transformations depended can be conceived of as acts of intelligence recognizing what to do with information available, and Newell and Simon eventually hit upon the idea of encoding each bit of knowledge in terms of a condition–action rule: *if* A *then do* B. This practice, along with the Newell–Simon–Shaw approach to structuring problem space, has become a standard feature of AI research (Waldrop 1988a,b).

The problem space with which Newell and Simon were concerned was of a kind appropriate for a *general* problem solver, accommodating both problems for which there is one and only one solution and problems for which there may be several solutions, some better than others. However, while their General Problem Solver was good at solving *some* problems, it never learned, on the basis of its own experience, whether it had done a good or a bad job (a serious problem, in terms of the present discussion of internal (i.e. 'self'-) communication. In effect, all it remembered was the

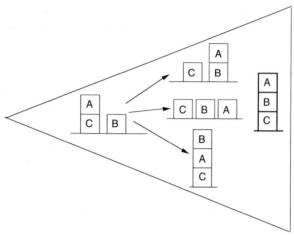

Figure 4.4.   Blocks world problem space. Starting from an initial state – the configuration of blocks on the left – the problem solver has three operators which it can apply to change the blocks to a new state: move block A onto block B, move block A to the floor, and move block B onto block A. The task is to find a sequence of such operators which will produce the stack ABC shown on the right. From Waldrop (1988a).

particular series of events leading to a solution and its own cry of Eureka! There was no basis for the transfer of learning or generalizing from past solutions, so that its performance in every non-reiterative trial was 'characterized by false starts and dead ends' (Waldrop 1988a). Partly for that reason (according to Waldrop), partly as a consequence of changing rewards and incentives, many AI researchers gave up trying to build general problem solvers and turned to constructing 'expert system' programs. Expert system programs capitalize on the summed knowledge of specialists in, e.g., medicine, engineering, business, to compile a great number of if–then rules for addressing practical problems in one or another narrow domain of knowledge. The result was analytic problem-solving of great power within strictly limited domains of application, e.g., diagnosing disease, directing petroleum drilling operations, managing investment portfolios. The power of expert system programs in these highly specific areas overshadowed their limitations viewed from a research perspective, in which generalization of learning is the goal.

In recent years there has been a gradual resurgence of interest in general programs, due in part it would seem to marked advances in robotic engineering, in which one of the remaining major obstacles to development of, e.g., flexed pedal locomotion and highly dexterous, humanlike manipulation of materials, remains the construction of a cognitive system with common sense and tactical decision-making capacity – i.e., the ability to evaluate unanticipated situations and to decide for itself what information is relevant to the solution of problems posed by unusual circumstances. One way to compare a general program and an expert systems program would be in terms of condition–action rules and state configuration of working memory. A general problem solver may have relatively few condition–action rules, but it can fiddle away with those through a great number of intermediate states toward a solution. An expert system program has a far greater number of condition–action rules to work with but a correspondingly simpler architecture. An expert systems program does not need anything like the working memory of a general program; what it needs is a well-organized memory, an extensive series of catalogued, cross-referenced files, which are narrow but deep, into which a messenger with a map disappears and reappears with answers to questions.

What is needed, at least in the minds of those who believe that symbolic representation (and not, for instance, neural networking – see below) is an appropriate method for simulating human cognition, is some combination of general and expert systems programs, an approach to AI program construction that enables a system to ponder a problem in both its general and its particular aspects, moving around in a problem space which is well connected with a long-term memory of broad, not narrow, capacity. A truly unified approach to the simulation of cognition of course requires more than that. The simulated system must not spin its wheels if it arrives at what looks like an impasse, but must somehow get around that impasse and back on track to a solution; further, it must remember and in effect learn from what it has accomplished so that in future time and energy will not be wasted on altogether similar obstructions, including those which may be encountered en route to the solution of quite different problems (Waldrop 1988b).

Why a 'unified' theory, anyway? Not just because unification of theory is inherently a good thing, something that all scientists should strive for. In an extended argument for unified theories, Newell (1990) lists several reasons (see Box 4.1), the most important of which, apart from the fact that unification *is* a good thing, 'is that a single system (mind) produces all aspects of behavior' (p.17). Also, 'the mind is shaped by multiple constraints. It requires a unified theory of cognition to bring these constraints to bear, to discover how they jointly affect and determine the structure of mind' (p.18).

Box 4.1. Why unified theories of cognition are needed. (From Newell 1990.)

1. Unification is always an aim of science
2. Bring all parts to bear on a behaviour
3. Bring multiple constraints to bear
4. Increased rate of cumulation
5. Increased identifiability
6. Amortization of theoretical constructs
7. Open the way to applications
8. Change from discriminative to approximative style
9. Solve the irrelevant-specification problem

The constraints that shape mind are listed as follows:

1. Flexible behaviour as a function of the environment
2. Exhibition of adaptive (rational, goal-oriented) behaviour
3. Operation in real time
4. Operation in a rich, complex, detailed environment
   Perception of an immense amount of changing detail
   Use of vast amounts of knowledge
   Control over a motor system of many degrees of freedom
5. Using symbols and abstractions
6. Using language, both natural and artificial
7. Learning from the environment and from experience
8. Acquisition of capabilities through development
9. Autonomous operation within a social community
10. Being self-aware and having a sense of self
11. Being realizable as a neural system
12. Being constructable by an embryological growth process
13. Arising through evolution. (After Newell 1990:19)

It is important to note that for the most part they apply to the monkey or ape mind as to our own. Although constraints 5, 6, and 10 have long been viewed as unique to human cognition, few primatologists today would deny to chimpanzees at least a reflexive consciousness or the ability to use symbols and abstractions. Use of language, involving syntax, seems to be another matter (Premack 1986), and there is no denying that human beings, apes, and monkeys differ from one another with regard to the particular action exercised by many of these constraints. The point is that AI models of cognition, which for reasons having to do with the history of their development have aimed primarily at simulating human cognition, are more broadly applicable. The uniformity of their aim has less to do with peculiarities of human cognition than with the goals of research in human-oriented disciplines, with practical applications anticipated in commerce and industry (including not least industries important to space exploration and warfare), and, consequently, with grant funding encouragement (always an important shaping constraint on the human scientific mind!) They seem to us to provide potentially very useful tools, hitherto under-utilized, for examining not just the nature of the human mind but of primate cognition in general, especially perhaps from an evolutionary standpoint. It is for that reason that we draw attention to them here, and to Newell as exponent.

According to Newell, unified theories of cognition embodied in working computer programs are 'within reach'. Soar, the program on which Newell and his associates have been working since the early 1980s (Laird *et al.* 1986; Newell 1990), is one prototype of a unified approach. In it, they return to the production system architecture of the General Problem Solver, introducing two important innovations. The first has to do with conflict resolution when a number of different condition–action rules present themselves as candidates for the operant role in effecting transition from one state to another: a decision routine here insures consideration and comparative evaluation of all candidates. Still there might remain multiple candidates with equally good credentials. This impasse situation, as well as a similar one in which no condition-action rule seems to apply to the situation at hand, is handled by a mechanism called 'universal subgoaling' (Figure 4.5). Each impasse is treated in the same general way – a new problem space is constructed around it and the entire problem-solving apparatus of the program is brought to bear on a solution (Laird *et al.* 1986; Waldrop 1988a; Newell 1990).

Laird's initial work dealt primarily with impasses, Rosenblum's with the problem of how to program learning as performance improved through practice (Laird *et al.* 1986). Rosenblum introduced the familiar notion that information, however it may be retained in memory (that is, in whatever form) is retained in organized, activity-relevant 'chunks'. Chunking, in its processual application to learning, takes the empirical finding that performance at most tasks improves over practice trials and

LIVERPOOL JOHN MOORES UNIVERSITY
LEARNING SERVICES

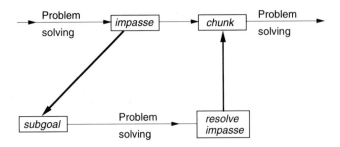

Figure 4.5.   When Soar reaches an impasse in its problem-solving – that is, when it does not know what to do next – it automatically sets up a subgoal to resolve the impasse. When it succeeds, it goes back to where it left off and simultaneously encodes a new 'chunk' of knowledge that will keep it from ever having to suffer that particular impasse again. From Waldrop (1988a).

applies it where appropriate to the formation of condition-action rules; it '. . . takes a complex, slow piece of problem-solving and replaces it with a simple, fast stimulus response reflex: "IF *this* is the situation, THEN do *that*"' (Waldrop 1988b, author's emphases; our summary to this point of the workings of Soar owes much to Waldrop's concise outline). Laird and Rosenblum recognized the importance to universal subgoaling of a similarly universal learning routine, and after completing separate dissertation projects (under Newell's supervision) they sat down and fused chunking and subgoaling to produce a program that could solve a difficult problem and learn something in the process. This it can be argued is exactly what is needed, to begin with at least, for a simulation of cognition to be interesting in terms of cognitive *process*. Basically, as the diagram in Figure 4.5 suggests, subgoaling enriched by chunking allows a working program to construct its own expert system subprogram *with respect to a particular problem of performance* and to encode results in the main program in the form of a condition–action rule. Returning to work on the larger problem it can clear the subprogram of work engendered by what at first presented itself as an impasse (the virtue of a general approach), in future calling into play the newly learned rule whenever it encounters a similar situation, whether in a rerun of the same problem differently configured or while grappling with some new task similar only with respect to this particular feature.

Of course, while this new generation problem solver may not be so easily baffled by what it conceives to be an impasse, it can arrive at the *wrong* solution, learn the *wrong* lesson from all that work. As Newell (1990) puts it, 'Chunking adds productions permanently to Soar's long-

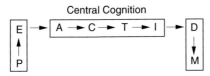

Figure 4.6. The minimal complete functional path from stimulus to response goes from perception (P) to encoding (E) to attending (A) to comprehending (C) to tasking (T) to intending (I) to decoding (D) to motor action (M). (From Newell 1990.)

term memory – it adds permanent structure. So now suppose that a bunch of these productions are wrong. What is Soar to do about it?' (p.466). The architecture of Soar, Newell suggests, 'contains two error-recovery mechanisms. One of them consists in the fact that Soar does everything in problem spaces: To go down one path in a problem space is to be capable of backing off, for whatever reason, and going down an alternative path' (p.467). However, if this were Soar's only error-recovery mechanism it might spend most of its time in blind alleys, repeating old mistakes on its way to each new solution. The second Soar mechanism for error recovery Newell calls the 'decision-time override'. This is a decision cycle mechanism which can operate, e.g., at the onset of Stage I Intend (Figure 4.6), in the pathway from perception to response, to forestall a wrong response.

A bit of explanation is in order here, and we will give it as far as we can in Newell's own words (1990). '*Intend* is the function of committing to a response . . . not the response itself . . . but the issuance of a command to the motor system that commits the system to a particular response' (p.265) – consequent on whatever may be necessary in the way of procedural review of past productions (as, in Figure 4.4, stacking blocks C, B, and A in order from bottom up). 'In the decision cycle, during the elaboration phase, everything pours in from recognition memory (the production memory) until quiescence occurs, after which the decision procedure determines the choice that will occur (including impasses). Thus, productions propose, decide disposes' (p.468).

Anything proposed by one production may be rejected by another. Thus, no single production can gain complete control over Soar's behaviour, and with the decision override there is always the opportunity to reject proposals of 'rogue' productions. However, such rejection of course depends on Soar's ability to learn new productions that will reject old, bad proposals that may seem to have served in the past. Techniques to do this have been developed (Laird 1988), but:

> Even with these techniques, using the decision-time override cannot be the complete solution. The elaboration phase would keep building up.

An organism that exists over a long period of time would steadily accumulate productions that propose actions along with the search control to negate their proposal. Each elaboration phase would have to recapitulate all of the bad learning that has ever been learned in that given context. Thus, these mechanisms do not seem sufficient, just by themselves, to solve the problem of recovery from a bad structure.' (Newell 1990:468)

Newell goes on to discuss further possible approaches to a solution, but we can leave the discussion at this point. The goal, which appears to involve perfecting the evolution of cognition in future machine generations, is not irrelevant to our concerns, but what interests us more is the fact that Soar at present appears to share the fallibility of living organisms. It is true that human beings and other animals are capable of recognizing when they have made a mistake and of reformulating behaviour so that they do not make the same mistake again. But it is true also, is it not, that individuals in their lifetime, and evolving species, are quite capable not just of repeating grievous errors but of piling them up in series with catastrophic results.

Soar is not the only example of an AI program which can apply general reasoning techniques to complex intellectual problems and learn from experience in solving them, but it seems to us to embody best that unified approach for which Newell is such a strong advocate. Not everyone agrees with Newell that unified theories of cognition are within reach and worth working toward. Connectionist and neural networking critics of symbolic representational versions of AI argue that 'the idea of producing a formal, atomistic theory of the everyday commonsense world and of representing that theory in a symbolic manipulator' is empirically untenable (Dreyfus & Dreyfus, 1988). Neural networking, an alternative approach to simulating cognition (more narrowly defined as neural activity associated with pattern recognition) may provide a better model of how the brain performs as a physical system. The debate is one which we will not pursue. We have discussed Soar at some length because, whether or not Newell is correct in his projection of ultimate success, this attempt to embody a unified theory of cognition in a working model appears to us instructive not only with respect to the ontogeny of learning and the acquisition of problem-solving skills by individuals in development but with respect to the *evolution* of learning and of principled decision-making. Implications for the evolution of human culture we leave for consideration after we have reviewed primate cognition from a more traditional perspective and have examined findings from recent field studies of primate social cognition.

In their general character – to anticipate a bit – those implications seem clear. Basically, the AI approach raises interesting questions about how working memory is related to long-term memory and how that relationship has evolved; how decision–action rules that govern tactical decision-making may be discovered or formulated in experience and whether common experience must give rise to common discoveries; how information is 'read into' internalized expert systems programming; whether animals lacking human language may nevertheless have access to extra-somatic expert systems (see our discussion of 'detached receiver/transmittors' in section 6); and, indeed, numerous questions concerning how information may be stored and accessed exosomatically, particularly by individuals in groups.

8. Processing social information

So far in this chapter we have introduced two important topics, the *institutionalization of social relationships* (section 4) and *information as a central feature of communication* (sections 5 and 6), both of which entail conceptualizations of behaviour which may be problematic where non-human primates are concerned but which appear to be essential to consideration of the evolution of human cognition and human culture. We propose now to beg questions such as whether information is a 'substantive attribute' as well as a 'central feature' of communication, and whether any signal can be said to have content apart from the formal relationships which make a pattern discernible. If readers can be satisfied that it is useful to conceive of information as being transmitted via signals – and subsequently stored, retrieved, and variously manipulated – then we can safely leave questions of essential character to philosophers. In this section we introduce a third topic which is currently of keen interest to primate behaviourists, and that is *social knowledge*.

It is difficult to think how the institutionalization of relationships might proceed without change in the character of social knowledge, and it seems to us equally difficult to discuss how the two are related without employing informational terms and examining the informational character of cognition and communication. However, we leave detailed discussion of these matters for subsequent chapters; those chapters, beginning with Chapter 5, are organized in a large part around just such discussion. Our object in this section is primarily definitional: what is meant by *social knowledge*, and what do we have in mind when we refer (as we did just above) to 'information stored and accessed exosomatically by individuals in groups?' Presumably this would be a store held in common by a group or community of groups, but would it be a store to which all members have *equal* access? A store to which all or any individuals have *full* access? *Continuous* access?

It may help to address such questions if we make a distinction between the *domain of social knowledge* (as discussed by e.g., Cheney *et al.* 1986 and Cheney & Seyfarth 1990) and the *social domain(s) of knowledge*, or *knowledge in a social domain*. The *domain of social knowledge* would be all of that knowledge having to do with social behaviour, from recognizing individuals (including oneself) *as* individuals to understanding the conventions which underlie society. Most of the current interest in social knowledge and social cognition revolves around attempts to contrast the domain of 'social' knowledge with that of 'object' knowledge. The contrast, implicit in a characterization of lemur intelligence by Jolly (1966), was fully drawn by Humphrey (1976) in a statement on the social function of intellect which deserves quoting at length:

> [O]ne generalization can I think be made with certainty: the life of social animals is highly problematical. In a complex society, such as those we know exist in higher primates, there are benefits to be gained for each individual member both from preserving the overall structure of the group and at the same time from exploiting and out-manoeuvring others within it (see later). Thus social primates are required by the very nature of the system they create and maintain to be calculating beings; they must be able to calculate the consequences of their own behaviour, to calculate the likely behaviour of others, to calculate the balance of advantage and loss – and all this in a context where the evidence on which their calculations are based is ephemeral, ambiguous and liable to change, not least as a consequence of their own actions. *In such a situation, 'social skill' goes hand in hand with intellect, and here at last the intellectual faculties required are of the highest order.* The game of social plot and counter-plot cannot be played merely on the basis of accumulated knowledge, any more than can a game of chess. (p.309, emphasis added)

It would be difficult to improve on Humphrey's discussion (above and in the 'see later' continuation) of how social forms emerge from the interactive exchange between individuals bent on using others to their own advantage – capitalizing on the behaviour of those others in cooperative as well as competitive manipulations – and of course it is clear in his argument that 'manipulation' is a feeble term for what is going on. Humphrey, as we note elsewhere, excludes information very usefully from his transactional view of social behaviour, and this paper is hence nicely complementary to those of Dawkins & Krebs (1978; Krebs & Dawkins 1984).

For it can of course be useful to examine processes of communication without reference to informational content; what is not useful is to suppose that because analysis of communication is furthered in some respects by leaving information out of account, the concept of information is therefore not helpful to the analysis of communication *in evolution*. One need not attend to the informational aspects of communication in discussing social cognition and *the domain of social knowledge* where concern is with how individuals use their knowledge of individuals, relationships, and group processes to manipulate both individuals and relationships. On the other hand, one cannot very well discuss individual cognition and *the social domain of knowledge* without attending to the informational aspects of communication, because the *social domain of knowledge* as we would define it is a supra-individual domain either exosomatic (as in an encyclopedia, software dictionary, or videotape) or distributed among several individuals (as in a council of elders or the group as a whole) – which is ordinarily, in either case, available in its entirety to few individuals and perhaps not to any one individual on a continuous basis.

Whether knowledge distributed throughout a social group is ever available in its entirety to one or more members is a nice question. A hamadryas baboon group just down from its sleeping cliff, extending and retracting its amoeba-like pseudopods in Kummer's (1971) description of pre-foraging 'debate' seems to be reviewing knowledge in the social domain. So, obviously, is a village council meeting to discuss a request for licensing for sale of alcoholic beverages. So too, less obviously, are tribal participants in (and onlookers at) a ritual dance, whether in Africa, North America, or Polynesia. Clearly of interest to studies of cognition (to say nothing of communication) in evolution are questions of how knowledge in the social domain is accessed, reviewed, updated, and restored (consider a dance or a song which is interrupted by a dancer or singer who, perceiving what he or she claims is an error in performance/copy, objects: 'no no, that's not the way it goes, it goes like this . . .' (see our example in Chapter 8).

Having examined such issues, we are tempted to conclude that *culture is largely a matter of processing information in the social domain*, and indeed that is how we shall now define it, extending the definition given earlier. Certainly it is hard to imagine how to think about social cognition *in its evolutionary aspects* without taking into consideration knowledge in the social domain, and there seems to us no way to conceive of knowledge in the social domain except in terms of information.

We can now return to Hinde's notion of the *institutionalization of relationships*. What institutionalization in general connotes (dictionary definitions are not much help here) is the establishment or implementation of anything on a regular and formal basis, formal at least in the sense of clarification of context and form so that agreement can be reached concerning component features. Definition of 'institutionalization' is important for our purposes only insofar as it sheds light on processes at work. 'Institutionalization of social relationships' should refer in some part to the development of structures and processes which increase agreement within a community as to what the respective roles and responsibilities of individuals to one another are (what constitutes a relationship) and what measures may be adopted to enforce performance – a court of opinion, then, as to whether a relationship exists and what sanctions may be applied against irresponsible action. Agreement seems to depend on 1. objective recognition of one's self and the selves of others as individual entities between which relationships can be objectively measured, and (2) communication within a group or culture about the character of those objectively observed entities and relationships.

If we were interested in 'social cognition' and 'social knowledge' only as these relate to contemporary non-human primate species, we could perhaps dispense with discussion of institutionalized relationships. However, as we intend to concern ourselves with the evolution of primate social cognition, we cannot dispense with it. Similar concern for the evolution of culture requires that we pay close attention to forms and processing of information. In our definition of culture, which is admittedly a comprehensive definition (and therefore, we believe, useful for making cross-specific comparisons), the focus is on social behaviour as a means of filing, accessing, and updating information in the social domain which is useful to individuals engaged in maximizing reproductive fitness. By this definition, it would seem, understanding the evolution of culture is essential to understanding the evolution of a characteristically human communication system (i.e., language), and not vice-versa.

Every human language, it has often been noted, is also a metalanguage which enables discussion of linguistic and other forms of communication. Thus the objectification of behaviour and of social relationships is concretely realized in the evolution of human communication. To be a language-using human being is to think in terms of role expression and of communication as revelatory social discourse. This does not mean that internal or external representation of an objectively realized self cannot be achieved without something like human language (cf. Bickerton 1990); it does mean that human beings must have a hard time thinking about how such representation might be accomplished.

9. Can a general theory be fashioned

Turning from causal explanations of behaviour associated with selfish-gene theory to a descriptive model (Hinde's) of how social behaviour appears to be organized in groups of individuals, we have focused on dyadic relationships between individuals. Focus on relationships is basic to summary description and analysis of social behaviour within and across age, sex and other classes and, to some extent, to the task of operationally defining social behaviour prior to collecting data. Hinde suggests (e.g., 1987) that a major difference between non-human and human social life lies in the institutionalization of relationships. Anthropologists may argue about whether kinship, for instance, is an entity in itself or merely a way of talking about customary associations, marital expectations, property rights, etc., but most will agree that in human evolution relationships became objectified so that parties involved can, so to speak, get outside their relationship in order to review and discuss with one another what seems to be required in the way of performance, which behaviours may be appropriate and which inappropriate, to what extent they may be reciprocal or complementary, etc., etc. We assume that non-human primates do not engage in such discussion, but the extent to which they may objectify a relationship and vary performance in ways which could properly be described as role playing is as yet not at all clear. Reflexive self-recognition would seem to be basic to the process, but the more one thinks about what that must involve the more difficult it becomes to define. Certainly that most popular operational measure of self-recognition, i.e., mirror recognition, is far from satisfactory (see Mitchell 1993 for review).

Selfish-gene theory tends to reduce group social behaviour to a lonely struggle between interacting individuals – one against all, with a little help from kin (Hamilton 1964) and friends (Trivers 1971). Hinde's descriptive model of the organization of this behaviour clarifies how interactions between particular individuals may be linked day-by-day in series characterized by meaningful patterns of continuity with changes to form relationships, with the implication at least being that the same processes (e.g., signal monitoring, feedback, and cognitive review) which underlie continuity of interaction may also explain the integration of relationships. Social organization, that is, may be fully explained on the basis of individual behaviour, in terms of genetic predisposition and of proximate constraints relating to perception, cognition, motivation, and of course anatomical and physiological conditions. This is not to suggest that it must be necessarily unprofitable to consider groups as entities in themselves, only that the character of group social organization – including institutionalization of relationships in human evolution (Hinde 1987) – is probably best analysed within a framework of theories focused specifically on individual behaviour, on the sociogenetic logic and cognitive underpinnings of decision-making, on dyadic social interactions and relationships, on behaviour in development, and, more generally, on an individual-oriented behavioural ecology and ecological psychology.

While we cannot expect to reconstruct in detail the evolutionary path which led to institutionalization of relationships for *H. sapiens* and perhaps not for other contemporary primate species, it seems clear that a general and unified model of cognition and learning would be essential to understanding the direction taken and the character and timing of events. If selection works to achieve a change from behaviour which is primarily *rule-governed* (i.e., obedient to recurrent physiological and external environmental proximate settings), to action that is *convention-governed* (guided more loosely by tradition and by institutional forms), then what selection must work on is individual cognition and learning. In the next chapter, therefore, we review briefly some basic and general aspects of primate cognition. This will provide the foundation for an evolutionary analysis of social cognition and social action in chapters to follow. It is, after all, the evolution of social behaviour in which we, as anthropologists, are particularly interested.

Notes

1. We recognize that cognition, especially in its reflexive character, is a concept which many scientists regard as highly problematic, especially perhaps those (e.g. some neurophysiologists and probably most psychologists and engineers engaged in research on pattern recognition) whose attention is not focused on whole organism behaviour and on the perspective of the individual *per se.*

2. See P.C. Reynolds (1981) for a thoroughly detailed, speculative but closely reasoned account of human language as a system of transactions which evolved from prior systems of social and material exchange. In this account no particular emphasis is placed on the informational elements of transactional activities such as grooming or manipulation of objects in a social context; it is the transactions *per se* which are important. Similarly, Humphrey (1976) emphasized the manipulative and transactional elements of primate social behaviour without attending to informational features of communication.

5   *Cognition*

1. Introduction

Social cognition, social intelligence, and social knowledge (Jolly 1966; Humphrey 1976; Smuts *et al.* 1987; Essock-Vitale & Seyfarth 1987; Stammbach 1988) are concepts important to understanding the meaning of primate social behaviour, both from the standpoint of behaving individuals and from that of anthropologists trying to utilize their observations of non-human primate life in explanations of the evolution of human social behaviour and its cultural bases. But these terms are sometimes used as buzzwords, and even when they are not there seems to be a tendency to blur distinctions between them. Hence this short review of cognition.

There is a second reason for pausing to take a closer look at cognition *per se*. If we are to make useful distinctions between social cognition, social intelligence, and social knowledge it is important that we first distinguish, if we can, between social cognition and cognition in its more general aspects, and between social intelligence and general intelligence.

We have yet a third reason for surveying the bases of cognition. Most important of all, it has to do with the informational approach we have taken to analysing social behaviour in evolution. Our concern is not with the physiology of perception and cognition but with ecological context. In subsequent chapters we will compare human social action with that of non-human primates in terms of information bases and information processing. It is essential that we begin with description and comparison of information processing at its most basic level – that of sensation, perception, and subsequent cognitive integration by individual primates. Indeed, most comparisons of animal intelligence (i.e., of general intelligence) have been conducted at this level, with an emphasis, usually, on ability to learn (Thomas 1980, 1986) and on 'a continuum of complexity rather than any dramatic difference in kind that might separate . . . our species . . . [from others]' (J. L. Gould & G. C. Gould 1986.)

Things in nature share features of elementary structure, so that boundaries tend to blur or grow fuzzy at one or another definitional focus. Fuzzy boundaries are not infrequent between taxa of natural taxonomies.

117

However, if categories of natural phenomena are naturally fuzzy, it nevertheless seems to most of us most of the time that things and organisms themselves do not grade insensibly into one another, that there is space between them, and that if the space is not exactly empty it is not filled with connective tissue either. Similarly, it does not take a mystical turn of mind to perceive that for mobile organisms making their way from one package of food to the next it is useful to perceive the world as broken into discrete entities separated from one another in different ways and to different degrees. The question is not whether we are fated to see as discrete and heterogeneous a world that is, if not homogeneous, nevertheless one organically connected whole – much as proponents of the Gaia hypothesis in its narrow version have suggested – but to what extent selection for discriminative perception, beginning with selective sensory input, may have encouraged the internal steersman of our sort of genetic vehicle (Dawkins 1976) to take such a view.

In order to examine individual behaviour at the environmental interface we must take into account the sensory modalities (vision, hearing, chemoreception, proprioception), psychological aspects of perception, and, more generally, cognition and learning. Our account of these will be brief; for a more detailed account of the anatomical bases of primate cognition see Martin (1990); for overviews of perception and cognition in psychological perspective see contributions to Fobes & King (1982a).

To speak of individuals in their relation to the environment is to suggest a relationship which is highly specific and at the same time rather ambiguous. How are we led to think about the interface? We may think about it for some purposes in terms of occasional exchange between an individual *and* the external environment (Figure 5.1a), for other purposes in terms of more continuous constraints on behaviour of an individual *in* an environment (Figure 5.1b); we may think of individuals as material entities in an environmental relation to one another (Figure 5.1c), or, as we are more accustomed to do, in terms of their social and interactive properties, (Figure 5.1d) and, finally, we may consider the social environment itself in relation to the organic and physical components of a habitat (Figure 5.1e) to which the term *environment* is usually referred.

These distinctions and the diagrams in Figure 5.1 may seem excessively formal for describing a relationship which appears to be, after all, simple in essence. But, as we have just argued, the relationship between an individual and the environment is not at all simple. The diagrams in Figure 5.1 constitute one of many possible sets of *visual* metaphors for this relationship; we include them partly for that exemplary purpose,

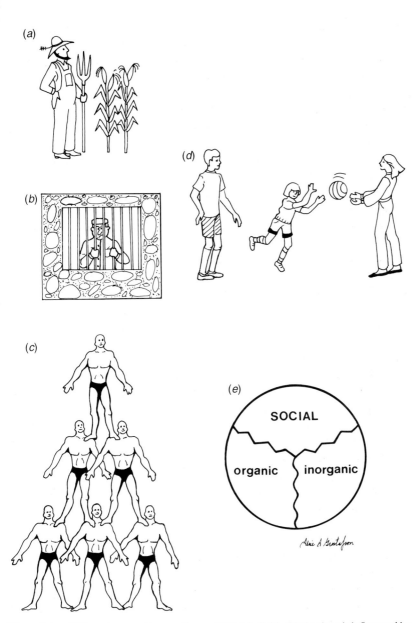

Figure 5.1. Different perspectives on the way which individual behaviour is influenced by the environment. (a) An individual conducting transactions with the surrounding environment; (b) An individual constrained by his environment; (c) A physical environment composed of individuals; (d) Individuals interacting in a reproductive and social environment; (e) Components of the constraining environment: overlapping and interactive in their influence on individuals. See text for discussion.

LIVERPOOL
JOHN MOORES UNIVERSITY
AVRIL ROBARTS LRC
TEL. 0151 231 4022

partly to remind the reader that society or some portion of it (the local group, family, household, etc., where human society is concerned), may be, and usually is, viewed as exercising a buffering influence. If this is in fact the case, rather than (again) just a way of looking at things, then it must have important implications for cognition – with different cognitive outcomes produced, for example, by mother–offspring relations, which are variable in character across species, or by different structures of close day-by-day association which go beyond the mother but which are still confined to sub-groupings within the larger group.

Robert Hinde's view of society as an organization of behaviour made up in individual interactions forming relationships nicely accommodates modern theories of individual competition and selection (Chapter 4). Such a view can help us address the question of how animals manage their relations with others towards the objective of greater reproductive fitness for themselves. And transactional (not necessarily informational) analysis (as in Chadwick-Jones 1987b) of the management of relationships entails examination not just of the signal as manipulative display, or of the actors' motivations in relation to some inferred goal, but also of perception as selective attention to communicative stimuli. The concept of feedback (Figure 4.2) suggests that signal reception and cognitive redintegration is as important to transmitting as to receiving signals.

If, as seems to be the case, the study of social behaviour in evolution must be initiated at the level of individual interaction, with attention to the mechanisms and processes on which selection can be presumed to work, some of the most important of these mechanisms and processes must have to do with how social behaviour itself is perceived by the actors engaged in it. We will leave consideration of social cognition for our next chapter however, and concentrate for the moment on mechanisms associated with characteristic function in primates of the sensory modalities, on perception, and on general cognition and learning.

2. Sensory modalities

Comparative study of perception, cognition, and learning begins with an examination of the senses: seeing, hearing, smelling, tasting, touching. Methods of examination involve a variety of techniques for recording the sensitivity of subject animals to environmental stimuli. The relationship between the physical properties of the stimulus and the character of the response elicited can be evaluated by techniques collectively described as psychophysical (Fobes & King 1982c). How to interpret differences across species in terms of broadly defined cognition, and especially with respect to the comparative measurement of intelligence (see below), is a problem, but it is less of a problem within the order primates than it is across more distantly related groups of species. Description of differences and similarities within the order begins with the close comparison of two or more species; gradual extension of the base affords reliable comprehensive generalizations, which is to say that larger figures may be described without loss of essential detail (Passingham 1982). It is in these larger figures that the general pattern of primate evolution is best revealed, and we will attend primarily to them. In this overview of the senses we focus primarily on vision; then review briefly hearing and olfaction before turning to aspects of (again, primarily visual) perception. Tactile sensing is at least as important as these in primate life, but it has been so little studied from either an informational or a communications standpoint that we leave it out of account.

Vision

Primates have good visual acuity and colour vision, but then so do many other animals, e.g. birds. The character of vision varies so markedly between vertebrates and invertebrates – how for instance might one go about comparing the vision of primates (Martin 1990) with that of bees (von Frisch 1954, 1967) – and only less markedly between vertebrate groups, that the question must be 'good compared to what referent species?' Our reply is 'good compared to most other mammals'. As Fobes & King have noted:

> students are often surprised to learn that . . . mammals as a group have fairly poor vision compared with birds, reptiles, and fish. Good color vision is unusual in mammals, whereas it is prevalent in the latter groups. Primates are a clear exception . . . Many primates possess excellent color vision, and other types of visual sensitivity, such as acuity, are also well developed. (Fobes & King 1982c:219)

As primates we are perhaps less interested in the inferior vision of mammals generally than in our own ability to see well, but it is because of the background condition that that ability seems to require explanation. Mammalian poor vision can be attributed in a general way to phylogenetic inertia, specifically to nocturnal adaptations, ground foraging, and primary dependence therefore on smelling and hearing. The exceptional vision of primates appears to be related to foraging practices, particularly that of monkeys and apes engaging in *diurnal foraging in an arboreal habitat*. So far so good, but it may be difficult to bring our minds to bear on the central and holistic character of foraging for most animals, including the non-human primates. Human subsistence behaviour in agricultural and market economies is, after all, far removed from the day-by-day foraging of other mammals, which is best described for all but the human case as conducted by individuals, alone or in social groups. In marked contrast to such *individual foraging* is the *dual reciprocal subsistence system* of human hunter–gatherers, who practise regular division of labour and subsequent sharing of nutrients gleaned (exceptions which test this contrast are few and partial, e.g., hunting cooperation and food sharing by wolves, wild dogs, and chimpanzees). Certainly we think of locomotion as essential to foraging, but we nevertheless are likely to think of it also, where we ourselves are concerned, in conjunction with a host of other activities, most of which are only very remotely related to food procurement and feeding. It would be odd indeed if, asked to list the contexts of human locomotion, people habituated to modern urban life found their thoughts turning automatically to the procurement, processing, and ingestion of natural foods. We may locomote through the aisles of supermarkets, into the house with groceries, and up to the table at meal time, but for men at least (and not least male scientists) that is frequently as far as it goes. Eating 'on the run' (more likely in the car) for most of us is an exceptional and, in the sense that it has little to do with foraging *per se*, accidental conjunction of activities.

For non-human animals too, certain functions of locomotion, e.g., patrolling territories, seeking mates, transporting nesting materials, evading predators, have little or nothing to do with foraging *per se*, and much of the locomotion that *is* concerned directly with foraging may yet be perceived by observers as activity sequences distinct in terms of time, space, and apparent function from those devoted directly to extracting, processing, and ingesting foods. The emperor penguin, relieved by his or her mate from egg-tending and setting off on a trek of as many as 70 miles to the sea and food, is one of the best examples of such separation, but most land vertebrates move from feeding patch to feeding patch (that, after all, is what being a vertebrate animal is primarily about); individuals and groups even in prototypical grazing species such as wildebeest do not graze in an uninterrupted sweep across the land, and monkeys and apes likewise must move from tree to tree.

However, for primates feeding in nature, locomotor form, whether quadrupedal or bipedal – and, if the latter, whether bipedally striding or brachiating – seems to have at least as much to do with procurement processes as with getting to where the food is (Grand 1984). Until recently, it seems fair to say, anthropologists have been, albeit systematic, too intuitive in forging evolutionary links between primate vision and primate locomotion, object manipulation, grooming, etc., in their explanations of primate origins. If, as suggested by Cartmill (1974), the order had its origins in terminal end branch predation on insects, there are more completely integrated cross linkages between vision and locomotion, grasping manipulation, and feeding than most of us had imagined.

It is instructive to consider visual acuity and colour vision in relation to another aspect of primate vision – stereoscopy. Older interpretations (see, e.g., Le Gros Clark 1970), associate all three indiscriminately with a frugivorous diet, and, to put the case in terms which are not too oversimplified, assign to visual acuity the function of perceiving the outline of a piece of fruit from surrounding leaves, to colour vision the function of discriminating degrees of ripeness, and to stereoscopy the function of getting to it without miscalculating a leap and plunging to the forest floor. Compare this with Cartmill's prosimian model of primate origins (Figure 5.2), which relates visual and skeletal anatomic features to a diet of insects and which has the virtue of tying what is construed as an elementary pattern of prosimian behaviour to a presumed ancestral line of insectivores. According to Cartmill (1974) it was a foraging speciality, grasping predation of certain species of insects at the terminal ends of bushes and shrubs, which opened a niche for primate evolution. Food procurement by this means would have been most effectively governed by vision, a fact which has important implications for 1. rotation of eye orbits to a more frontal location, improving depth perception via stereoscopic overlap of visual fields, and 2. conversion of active behaviour to crepuscular and diurnal phases of activity. Daylight hours provide an advantage to visually oriented, not wholly arboreal small predators, an advantage not just in searching out and pursuing prey but in locating hunters for whom they themselves are the quarry.

Monkeys, apes, and human beings show improvements over prosimians in both visual acuity and colour vision, but, while it is easy to see the general benefits of improved vision to a visually oriented animal, especially to diurnally active primates inhabiting a complex three-dimensional world, one can only guess at which of those benefits if any may have served as over-riding 'targets' of selection. Indeed, it probably always involves serious oversimplification to think of evolution in such terms. It may be that development of primate colour vision went hand in hand with the evolution of dicotyledenous plants and fruits varying greatly in colour; on the other hand, it is unlikely that any species of primate, including the most insectivorous, views the world strictly in black and white and shades of grey. It is hard to estimate realistically how disadvantageous it might be for an old World monkey, ape, or early hominid not to perceive colour hues. Reported instances of absolute colour-blindness in humans are extremely rare; in a case described by Oliver Sacks (1987), psychological stress accompanying sudden onset of colour-blindness was extreme and long-lasting – and the case is unusual in that the patient was a painter by profession.

There is, in any event, close similarity among Old World monkeys, apes, and human beings in visual acuity, colour vision, and visual

Figure 5.2.   Cartmill's model of primate origins emphasizes grasping predation of insects at terminal branch ends in the understory.

behaviour in general, which suggests that divergent evolution of anthropoid visual systems peaked early on; subsequent evolution of vision, what there may have been of it, must have involved parallel adaptation (Fobes & King 1982c).

Hearing

The mammalian ear has a tripartite structure. The external and middle ear serve protective functions, the middle ear acting also as a receptor of auditory signals, while the inner ear is a transducer, transforming acoustic (external) energy into chemical excitation of sensory nerve fibers (Khanna & Tonndorf 1978). The inner ear is important not just for perception of sound vibrations but for perception of equilibrium and motion, which is perhaps a more ancient function (Bramblett 1976).

The ear, like the eye and the nose, is a *distance receptor*; it is especially useful for detecting predators and prey and pursuing a normal social life. In comparison with the eye and the nose, in terms of relative efficiency, what is perhaps most important is the relative velocity of sound (as of light) to locomotion. Smell is slower than either and is dependent on air movement. Light of course is faster than sound but is offset by certain disadvantages, being dependent both on adequate illumination and a direct, unobstructed pathway.

While an acoustic detector is potentially of great use, how well it functions depends on 1. sensitivity of detection, 2. omni-directional reception, with no dead zones, 3. ability to localize the source of sounds detected, 4. elimination of noise, and 5. identification of the sound source from analysis of acoustic properties. These properties in turn depend on numerous factors, e.g., for sensitivity of detection: minimization of transmission losses, increase in energy collection (gain), and sensitivity of the biological mechanodetectors involved; and, for directional reception of sound: interaural differences in the onset and intensity of signal reception (Khanna & Tonndorf 1978).

The ear, again like the eye and the nose, is an important *organ of communication*. Auditory aspects of vocal communication include 1. the auditory signal and 2. receivers' perception of vocalization. Research in both has been markedly advanced over the past two decades (for reviews see Eisenman 1978; Snowdon 1986; and, for the neural underpinnings of vocal production and vocal perception, Jürgens 1990). In the mid–1970s Marler (1973, 1975) argued that species of primates living in different habitats differed in the degree to which they exhibit calls which were discrete and calls which were continuously variable ('graded' or 'intergraded') in structure. Early studies of primate communication seemed to indicate that primates living in broad, open habitats, in which visual communication was largely unobstructed, tended to emit primarily graded vocalizations, while the repertoires of species in forested habitats appeared to contain a greater variety of discrete, categorizable calls. Marler's proposal triggered a flurry of research on both the structure and perception of primate vocalizations and generated close comparison of primate auditory communication systems with those of birds (Mitani & Marler 1989) and humans (Gordon 1990; Jürgens 1990), in both of which discrete vocalizations play an important role.

While Marler's generalization concerning open country versus forested habitats seems to hold for a number of primate species, its explanatory value is problematic. All primates hear well, and it is hard to generalize about particular advantages associated with the differences in competence which have been described. And, as Snowdon notes:

> [T]he distinction between discrete and continuous vocalizations has become blurred in recent years, in part through our knowledge of human speech perception. Several of our speech sounds would appear to form a continuous distribution when examined spectrographically, yet we rarely have difficulties distinguishing which category a sound falls into. This phenomenon of categorical perception of sounds that appear to be graded makes it difficult to apply the graded–discrete distinction to primate calls. (1986:499)

Nevertheless, subsequent rapid progress in the study of vocal communication, compared with that of communication in other modalities (visual, chemical, and tactile), appears to be due in part to concentration of research effort on puzzling out just how signals differ from one another, in physical structure and in receivers' perception. Understanding of both is essential if we are to determine the extent to which non-human primates are capable of communicating information about the external environment to one another.

Olfaction

Scents are important indicators of the physiological and sexual states of animals; they signal ripeness and edibility of plant foods; and, in general, they constitute for primates as for other mammals a direct and primary source of information about the external environment. Odour is an ancient component of mammalian communication (Keverne 1980; Jolly 1985), and while relatively few mammalian odours appear to be as unambiguous in signal content as are insect pheromones, olfaction nevertheless continues to serve critical functions in intraspecific communication where most primate species are concerned.

Scents are of particular importance, in the absence of light or in dim light, to animals of nocturnal and crepuscular habits. It is difficult to draw correlations between morphology and behaviour where olfaction is concerned; however, vomero-nasal and olfactory systems are relatively well-developed in prosimians and New World monkeys, and species in these groups appear to rely more on chemical communication as, more generally, on information processed through the olfactory system, than do Old World monkeys and apes (Epple & Moulton 1978).

In all primate species scents communicate information of various kinds about the individuals who produce them. The great majority of primate chemical signals received are not directed toward particular receivers as are many visual and auditory signals. They appear to serve primarily to broadcast the signal sender's species and individual identity (and, perhaps, group and/or lineage affiliations), sex, developmental age, and more transient conditions such as reproductive status, social status, and current emotional state. Alone or in combination with a call or visual display a chemical signal can fulfil different functions, serving as social or sexual attractant or as threat depending on context and on previous interactions between sender and receiver (Epple & Moulton 1978; Epple 1986).

The study of olfactory communication, as, indeed, of visual and vocal communication, is beset by many problems, including characteristically small sample sizes, highly variable responses, and lack of control over subjects' prior experiences. There is in addition to these a problem more fundamental: primate scents, except in a few clear cases of chemical control over sexual maturation and (in conjunction with external environmental triggering factors) seasonal sexual activity, 'do not function like insect pheromones that switch others' behaviour on and off. Primate scents, like other primate signals, seem to influence rather than control the receiver's actions.' (Jolly 1985:203).

Study of chemical signals is further complicated by the fact that, in comparison with vocal and visual signals, they are difficult to record and preserve. Nevertheless, once they have been collected, mass spectrography and gas chromatography enable accurate analysis; also, where chemical signals can be synthesized and emitted in timed and graduated quantities to test subject animals' responses, experimental control is greatly increased. According to Snowdon (1986), and largely for the reasons just noted, 'the study of chemical communication is second to the study of vocal signals in terms of . . . achievements to date' (p.495).

Epple (1986), with an eye to problems of data collection, says that the study of chemical communication, compared with that of the visual and auditory systems, is still in its infancy'. 'This is not surprising', she says,

> . . . considering the limitations placed on the human observer by his or her own sensory capacities. Because of a relatively poorly developed sense of smell, the human observer is much more likely to record a visual or vocal pattern than an odor, and much of the chemical communication that may occur in a primate group escapes notice during direct observation. (p.531)

She notes the improvements in analysis afforded by chromatographical and other techniques which enable researchers to isolate and identify suspected chemical communicants.

> However, great difficulties are encountered in separating signals from noise by means of studies aimed at documenting the behavioral relevance of chemical compounds . . . [T]his is partly due to the sometimes overwhelming chemical complexity of the signal–noise mixture, and partly to the ambiguity of the animals' responses to chemical signals.' (p.531)

Summary

The problem for human observers of discriminating non-human primate signals from noise and clarifying ambiguous responses to signals, whatever the sensory receptor system involved – visual, auditory, or chemical (we have left tactile signals out of account in this brief discussion) – is complicated by the interplay between modalities in most of the communicative interactions on which primate social life depends. One way of getting around that problem (but not of resolving it) has been to compare the several modalities in terms of communication system 'design features' and logical considerations relating to function and, more tenuously, to the evolution of behaviour (Hockett 1960; see also Hockett & Altmann 1968 and, for an informative summary, Bramblett 1976). For instance, one or another communications 'channel' may be more or less appropriate for *broadcast* or *directional transmission* (the olfactory channel is inappropriate for directional transmission, given the signal sender's lack of control over the vagaries of air currents – we leave subsequent comparisons to the reader), while signals may be transmitted at different power intensities or concentration levels, beamed over a limited range of frequencies, or even encoded against decipherment by alien receivers. Signals may constitute *addressed messages*, intended for a particular recipient. They may be subject to *rapid fading*, a virtue in many respects, the shortcomings of which may be remedied by repeated transmission, and characterized by either *straight line* or *crooked line transmission*, travelling direct from sender to receiver or, in the latter case, circumventing obstacles that would defeat straight line transmission (Bramblett 1976).

Comparison of these and other design features can improve our understanding of what constitutes a communication system and of the functional fit between communication subsystems (e.g., the visual, vocal–auditory, olfactory, and tactile channels) and environmental contexts of use. Comparison of design features so far does not seem to have improved by much our understanding of the evolution of non-human primate communication or, in particular, of human language. We leave further discussion of primate communication for Chapter 11, where we will approach it from the standpoint of language origins. Our main concern in these sections has been with vision, hearing, and olfaction not as communication systems *per se* but as information input systems. In each we have found considerable differences between prosimians and higher primates, not so much among Old World monkeys, apes, and human beings. At least, for these anthropoid primates, the bases of visual, auditory, and chemical perception appear to be generally equivalent. The *character* of perception nevertheless may vary greatly across and perhaps within species in response to ecological determinants.

3. Perception

Perception is not just a matter of detecting sensory stimuli. It is selective, interpretative, integrative, and is traditionally viewed as a complex of psychological mechanisms and processes intermediate between sensation on the one hand and cognition and learning on the other. Sensation is discriminative, but it is only in consideration of perceptual conditions and processes that sensory discriminations take on meaning. Music provides a classic example of the difference between sensation and perception: 'melody arises out of the relationship (perception) . . . [in a sequence of] tones rather than from any particular tone heard (sensation)' (Riesen 1982). A shift in tempo and introduction of harmonic changes and grace notes may alter the song, but the melody lingers on. An equally familiar example of the importance of context is the figure–ground phenomenon in visual perception, in which two vertical squiggles transform themselves from a vase in outline to facing profiles and back again (Figure 5.3); similarly, in the childhood puzzles of an older generation, the outsized likeness of an animal or person conceals itself in a bush or a leafy tree (Figure 5.4).

That perception may involve the application of multiple measures is evident in the way we combine information provided by the retinal image of an object with information concerning the distance from which the image is projected to arrive at judgements as to size. Misinformation about either can throw us off if, e.g., the projected image is of an object deceptively similar to one more familiar and of markedly different size, or if the image is of familiar objects projected via a lens-shaped mirror, as in automobiles of recent make in which a glimpse of one's own surprisingly small face evidences the need for the warning that 'Objects may be closer than they appear'. Size constancy depends on learning – in both of these instances confusion results from a confounding of familiar expectations – and feedback plays an important role: perception of distance itself seems to depend in large part on assessment of the configuration of recognizably familiar objects and of sizes predictable with comfortable reliability on the basis of experience. This must be of importance to arboreal primates, and to young monkeys in athletic play, alone or in rough and tumble

Figure 5.3.   Vase/profile.

Figure 5.4.   Face in tree.

games of chase, which characteristically repeat over and over leaps across the same gap and rapid traverses over a problematic section of substrate (Symons 1978). Whatever the motive underlying such play, an important outcome is confirmation of size constancy and practice at distance estimation.

For primates, perception of depth is improved by *stereopsis*, the superimposition of two different images in forward facing eyes. This advantage decreases as distance from objects increases, and even where it is pronounced to begin with, as in tarsier, it may be most advantageous in connection with prehensile operations involved in feeding, grooming, and the slower forms of quadrupedal locomotion. It has long been conventional in anthropological discussion of primate origins to empha-size the advantages of stereopsis for arboreal locomotion, but the emphasis usually has been on locomotion involving saltatory leaps – by monkeys and prosimians as well as apes (that is, by fast-moving branch-runners, hoppers and leapers, and brachiators alike.) By contrast, if the order is descended from more slowly moving quadrupeds accustomed to seeking out and pursuing insect prey in the small terminal branches of forest undergrowth, stereopsis may have had a more specific adaptive function relating to grasping predation. For the more social monkeys and apes, a complementary benefit of stereopsis, providing the basis for subsequent exaptation (Gould & Vrba 1982), could have been enhancement of reception of any visual signals in which information is embodied in a three-dimensional architecture – which presumably would include most if not all facial and postural–gestural signals – but about this we can do little more than speculate.

In theory, the perceptual advantage of stereopsis must be with regard to things in their horizontal aspect. Head-cocking behaviour (reviewed by C. B. Menzel 1980) may be construed as a compensatory behaviour evolved by primates, and particularly by smaller primates, to transpose that advantage to deal with things in their vertical aspect – but head-cocking in primates is a characteristic not simply of small animals but especially of young animals (Riesen 1982). Of primary importance to locomotion and feeding is estimating correctly the distance not between one and another object but between each of those objects and oneself. In a world of familiar objects this is probably effected most efficiently not by head-cocking and rotating the visual plane sideways but by eye movement that tilts it upward and downward.

Experience for human beings at least may influence emotional colouring of perception, in ways familiar and for the most part obvious. Colour preferences themselves are not simply a matter of learning, but for human beings colour perception is constrained by culture, by culturally varying uses of colour in art and other manufacture, in ways that ordinarily are taken for granted. We have noted the emotional impact of sudden, complete failure of colour vision, but absence of colour is clearly not disturbing in the context of, for instance, a darkened art cinema (though it may be less tolerable to a public conditioned to technicolour film and colour TV.)

Cross-modal transfer of perceptual learning
Object recognition, spatial perception, touch and haptic perception by primates have been much studied in recent years (for discussion of the complexities of haptic perception, including responses to body movement and manipulation of objects, as well as touch, see Riesen 1982), but

experimental psychologists' treatments of these subjects lie outside the scope of this brief survey. Transfer of perceptual learning across sensory modalities is of interest to students of primate communication and human language origins (Hewes 1973a), and primates vary considerably in their responses to tests of, especially, visual–haptic transfer of object discrimination. Monkeys do poorly on such tests (Ettlinger 1970); chimpanzees and orangutans and children aged three to five years do much better (cited by Davenport & Rogers 1970; Rudel & Teuber 1964; and Riesen 1982). Such transfer appears relevant to any consideration of the possibility of symbolic gestural communication preceding human glottogenesis, and so the difference between monkeys and apes is interesting, but in these and other comparative tests of performance it is difficult to be sure that tests are not biased in favour of apes and human children. At least as relevant to speculation concerning a gestural stage in language evolution (Hewes 1973a,b) is visual–auditory transfer of information. 'Temporal intervals, rhythmic patterns learned with sounds or light flashes, show transfer across the visual and auditory modalities better than does visual–haptic transfer of three-dimensional objects' (Riesen 1982:281), but whether this is a difference in the character of transfer mechanisms or of information involved is difficult to say. The aforementioned tests of visual–auditory transfer involve the embodiment of a common temporal pattern in different media; in tests of visual–haptic transfer, the same objects are exposed to different sensory modalities.

Social perception

Primates, like most if not all mammals, recognize one another as individuals, making distinctions on the basis of visual, vocal, and probably olfactory cues. More remarkable may be the ability of monkeys and apes to recognize individual social relationships and relational categories (Cheney & Seyfarth 1980, 1982a,b; Dasser 1988). Remarkable too is the ability of chimpanzees and orangutans (Gallup 1977), and probably gorillas (Patterson 1986; cf. Suarez and Gallup 1981), to recognize themselves in mirror images. It is important in both cases that we understand what is being measured and defined as 'recognition', for what is perhaps most remarkable is how difficult it appears to be to make reliable inferences concerning mental processes and conceptual abilities from observed overt behaviour. But, in any event, social perception and social cognition vary in degree across species: it appears to be useful to distinguish (a) recognizing other individuals as individuals from (b) recognizing their relationships to one another, and to oneself, and (c) recognizing oneself as an actor with distinguishable roles. It probably would be useful too to try to clarify precisely differences between social perception and social cognition, but at this point the social aspects of perception become problematic and interesting in ways which require more detailed treatment, so we leave further discussion for the next chapter.

4. Cognition and learning[1]

Until recently, the main goals of research in primate cognition were to discover the character of particular kinds of learning and to evaluate intelligence on a comparative basis. The basic method of that research was to measure an animal's performance, subject to experimental control, in tests of simple conditioning, of object discrimination, and of the ability to transfer appropriate discriminative reaction sets from one pair of test objects to another (Harlow 1944). Primates especially have been tested on problems more complex than these, problems which cannot be solved simply by responding appropriately and on time to a particular stimulus or set of stimuli; rather, in problems of complex learning, solution is likely to hinge on accurate perception of relationships between stimuli and configurations of elements in the test environment (King & Fobes 1982).

The continuum between simple and complex learning problems is best seen in tasks which involve clearly discrete choice but require discriminations in which complex rules determine which of sometimes several alternative responses may be correct. The rule is often conditional in that a correct response depends not on some property of the stimulus but on the context in which it occurs. An *oddity problem*, for example, may involve the presentation of a triangle in conjunction with two squares, the triangle to be chosen not because it is a triangle but because it appears once instead of twice. Two-odd oddity procedures utilize two identical pairs, e.g., two triangles and two squares, and a triangle in conjunction with two squares in equal frequency with a square in conjunction with two triangles. Since subjects can solve even two-odd problems by applying two separate rules ('if two squares, pick the triangle' and 'if two triangles pick the square'), a next step is to administer tests involving multiple sets of two-odd problems and stimuli pairs to which a subject trained in two-odd problems has not previously been exposed. Correct responses under these conditions presumably evidence mastery of a generalized 'oddity' principle, although there remains some question as to what rules may be used in applying this principle.

Sameness–Difference learning involves discrimination similar to that in oddity learning, but correct choices are determined by compound stimuli, e.g., colour and shape. Further complications can be introduced as in the Weigl oddity problem described here by King & Fobes:

> The Weigl oddity problem exemplifies application of the Weigl principle [referring to an ability to shift efficiently from one dimension to another in classifying objects of different forms and colors] to oddity problems with two possible dimensions in which to designate an odd object. Young and Harlow (1943a) trained rhesus monkeys on such Weigl oddity problems in which three objects appeared on each trial. One object was an odd color and the other was an odd shape . . . When the objects appeared on an orange stimulus tray, the odd-colored object was correct; when they appeared on a cream colored tray, the odd-shaped object was correct. Thus, tray color was the sign designating whether color or shape was the relevant dimension for oddity. Young and Harlow (1943b) later demonstrated that their rhesus monkeys could generalize Weigl oddity when more than three objects were simultaneously presented and when novel stimuli were used. (King & Fobes 1982:345)

Testing procedures involved in the Weigl oddity problem are representative of those in a large and varied group of *higher order sign problems*, designed to afford rigorous quantitative measures of success in dealing with complex concepts and generalizing learning across different situations. The extent to which primates or other animals learning to solve such problems may use concepts familiar to human beings is not at all clear. Attempts to study directly and with similar experimental rigour a subject animal's grasp of concepts such as number have not been particularly enlightening. In *matching-from-sample* testing (of figures of one sort to the same number of figures of another sort) chimpanzees have performed well at matching representations of one to seven figures (Ferster 1964; Ferster & Hammer 1966); on other tests (Thomas & Chase 1980; Thomas et al. 1980) squirrel monkeys appear to evidence as good a grasp of number. Rohles & Devine (1966, 1967) tested the ability of chimpanzees to learn to select the middle object in odd-numbered arrays, varying in number, in an arc of 25 foodwells on a stimulus tray. Initial training involved unbroken arrays, after which gaps of varying length were introduced, complicating the problem of bisection. Rohles and Devine's study of 'middleness', like other investigations of number-related concepts, provides no overt evidence for counting, but as the problem was complicated response latencies increased; this suggests 'operation of some type of vicarious trial and error. It would appear that

some type of counting must have mediated the chimpanzee's performance.' (King & Fobes 1982:351).

In an interesting study of hierarchical rank order (McGonigle &
Chalmers 1977), response latencies provide a similarly tantalizing clue to
concept manipulation in social behaviour. McGonigle and Chalmers
presented monkeys with nearest neighbour pairs drawn from an ordered
array of stimulus objects, i.e., stimulus objects 1 and 2, 2 and 3, 3 and 4,
etc., training them to respond to right-hand stimuli, i.e. stimulus object 2
in relation to 1, 3 in relation to 2, and so on. In a second stage of the
experiment subjects were trained to maintain their responses to particular stimulus objects whatever the right-hand–left-hand positioning in
paired presentation, e.g., 1 2 or 2 1, selecting stimulus object 2 in either
case. When in the third and final stage subjects were presented with
previously unmatched pairs, e.g. 3 1 and 2 5, there was an inverse ratio of
response latency to rank distance between paired objects; subjects were
likely to hesitate longer, for instance, before choosing 3 over 1 than 5 over
2. Whatever the implication here with regard to 'operation of some type
of vicarious trial and error' (see just above, King & Fobes 1982), that
hesitation appears to indicate an interesting application of transitive
reasoning in response to a hierarchy of stimuli. Similar application of such
a principle clearly would be of benefit in interacting with associates of
status variable in relation to one's own (Harcourt 1988, 1989). Minimiz-
ing risk in social interaction with a very high ranking individual may
depend in part on rapid if rough calculation of social distance, while
interaction with an individual of status appreciably lower than one's own
can proceed for all practical purposes without calculation. Careful
deliberation and thoughtful concern for the amenities may be most
important in dealing with those who are neither constant companions nor
miles away on the status ladder, individuals a step or two above or below
and likely to be sensitive to challenges intended or unintended. If
learning can be transferred from social settings to transitive relations
among material objects (Figure 5.5), then development in a hierarchically structured society would appear to provide subjects with good
training for tests like that of McGonigle and Chalmers.

Studies as rigorously controlled as these do not often show a direct
bearing on cognitive decisions 'in real life'. Experiments which allow
subjects more freedom of movement and/or regular interaction with
familiar associates may be more revealing in this respect. Menzel (1974),
interested in the ability of primates to solve everyday foraging problems
of distance, direction, and remembered location, conducted experiments
in which subjects' performances in retrieving food, e.g., from behind

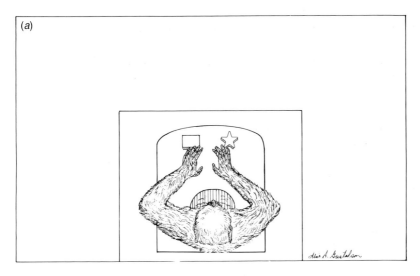

Figure 5.5.   Assessing hierarchical structure in (a) object arrangements and (b) social action. See discussion in text.

open-ended fence barriers, from experimenters, and parcelled out in 18 different locations, depended on their ability to employ more or less specific information about their habitat. In studies of forced delay in utilizing information about concealed food, subjects who watched via television monitor while food was placed in concealment by experimenters found it upon release with a degree of success intermediate between that of subjects allowed to watch directly through a window and subjects serving as uninformed control animals. This along with numerous similar applications of video techniques makes a convincing demonstration of the ability of primates to match two-dimensional representations with three-dimensional reality and to retain and apply information received via the former.

Notes

1. For discussion in this section we are especially indebted to Fobes and King's account of complex learning in primates (1982b); and we have profited throughout this chapter from our reading of the several excellent reviews of experimental research in primate cognition contributed to the collection in which that account appears (Fobes & King 1982a).

6　*Social cognition*

I have suggested that the life of the great apes and man may not require much in the way of practical invention, but it does depend critically on the possession of wide factual knowledge of practical technique and the nature of the habitat. Such knowledge can only be acquired in the context of a *social* community – a community which provides both a medium for the cultural transmission of information and a protective environment in which individual learning can occur. I propose that the chief role of creative intellect is to hold society together. N. Humphrey 1976, p.307 (Author's emphasis).

1. What is social cognition?

In the last few chapters we have reviewed strategic theories of behaviour (Chapter 3), discussed the tactics of individuals engaged in social interactions (Chapter 4), and examined the cognitive bases of learning and of rule-governed decision-making (Chapter 5). In this chapter we will consider what it is about the social behaviour of non-human primates that makes it not just interesting in its own right but important as well for understanding the evolution of complex intelligence. The evolution of cognitive abilities in primates has for many years been attributed to complex social organization and behaviour (Chance 1962). Chance developed his ideas in two stages. First, he emphasized the way primates constantly monitor each other in a social group, forming what he called an 'attention structure' for the group as a whole (Chance 1967). Later, he developed the theory of attention structure into a more complex argument that there are two modalities in social interaction, the agonic and the hedonic. The former gives rise to hierarchical social organization and authoritarian, non-innovate behaviour. The latter gives rise to egalitarian, innovative behaviour (Chance 1988). In Chance's view, each primate society, human and non-human alike, is characterized by more or less of these two features, and the proportion varies over historical time.

While primatologists as a group have not followed Chance in his bimodal analysis of primate society, the connection he postulated between primate cognition and primate social behaviour has come under close study. Most today probably would agree with Essock-Vitale &

Seyfarth (1987), that 'the intelligence of primates is most richly illustrated in their social life'. Of course, when animals engage in social activities that remind us of our own, as is so often the case with monkeys and apes, we may be inclined to perceive intelligence at work. But it does seem reasonable to suppose that if there is knowledge specific to the domain of social activity then primate social life may have provided the context for selection pressures of peculiar strength to operate on cognitive processes involving the manipulation of that knowledge. In primate evolution, it has been suggested, 'group life exerted strong selective pressure on the ability to form complex associations, reason by analogy, make transitive inferences, and predict the behaviour of fellow group members' (Cheney *et al.* 1986:1364). Accuracy in predicting the behaviour of conspecific companions must be important to animals engaged in maintaining, manipulating, and generally exploiting social relationships to their own advantage – and we hardly need add that being able to predict behaviour is at least as useful to cooperation as to competition.

The individual primate interacts with multiple partners in pursuit of a host of goals, short-term and long-term. The short-term goals are not always consistent with one another, nor is it always clear just how they bear on the over-riding ultimate goal of each individual to increase her or his contribution to the population's gene pool. The animals that we observe may be 'vehicles' of genetic adaptation (Dawkins 1982), but they are not *simply* vehicles. In considering how reproductive fitness may be maintained and replication of genes maximized we cannot leave individual action out of account; without due attention to the integration of life processes, body functions, and information processing at the *individual* level of behaviour organization, it is difficult to see how developmental and evolutionary variety can be fitted together in a comprehensive explanation of behaviour under selection.

How adaptation occurs and *what* adaptations occur depend in part on the way in which individuals engaging in social action are accustomed to organize their communicative behaviour. Concern for that organization has stimulated interest in the study of primate social cognition, or of general cognitive processes as they apply to social behaviour. Dunbar (1988) has remarked upon a shift of perspective in recent years 'away from an essentially behaviourist perception of animals as genetically programmed machines . . . to a more cognitive view in which animals are conceived as evaluating strategies in a rational way' (p.323).

Central to this 'more cognitive view' is the assumption of intelligence and thoughtfulness in non-human primates, qualities which however difficult to define we begin by assuming, in the case of apes and probably

Old World monkeys, ought to be dealt with in terms similar to those we use in describing our own behaviour. A working definition of social cognition in human beings and apes might be *The application of intelligence to the review of social information and the exploitation and management of social relationships toward attainment of short- and long-term goals.* The emergence of social cognition as a topic of concern for research in primate behaviour does not mean that studies of primate social cognition will closely resemble studies of human social cognition. In human psychology, where research in social cognition is a well-established tradition, social information is taken for granted in the same way as are intentionality and purpose in human behaviour. Taking it for granted, students of human social cognition are not interested so much in the character of social information, at least not in establishing what is distinctive about it (cf. Humphrey 1976; Cheyney *et al.* 1986; Crook 1989) as they are in the ways that perception may be influenced by internal values, needs, and expectancies unrelated to stimulus qualities (see, e.g., Postman *et al.* 1948, and Higgins & Bargh 1987). Typically, research into human social cognition has involved asking how it is that people do not always or even typically 'reason about social information by weighting and combining evidence in a rational and judicious manner' (Higgins & Bargh 1987; and see also Heider 1958; Mischel 1968; and Kahneman & Tversky 1973, all cited by Higgins & Bargh.) Psychologists who study human social cognition have tended to focus on 'ways in which people go *beyond* the [social] information given' (Higgins & Bargh 1987).

Researchers in primate social cognition are more interested at this point in *what constitutes social information* and, as indicated by the quotation by Nicholas Humphrey that opens this chapter, in asking what evolutionary relationships there may be, if any, between social cognition and the processing of behaviour in apparently non-social domains (e.g., in the use or manufacture of tools). We note this difference of focus between research into the social cognition of human beings and that of non-human animals partly to point out that consideration of work in the first area has so far been almost entirely omitted from discussion of the second. Considering how much attention is devoted in the study of human social cognition to manifestations of bias, irrationality, and 'going beyond' information given, this is an important omission. Comparing the two bodies of literature one is led to wonder whether we have not tended to attribute to non-human primates behaviour that is more rational, in situations where fitness is at issue, than is that of human beings. It may be, of course, that primates whose minds are free of philosophical uncertainties make better fitness logicians, but the question is open, nor do we know for that matter what existential anxieties may be entertained by monkeys and apes.

2. The uses of social knowledge

There is clear evidence that monkeys, apes, and other primates recognize one another as individuals, as apparently do other mammals, birds, and perhaps most vertebrates. What is at least as interesting is that monkeys and apes appear also to recognize relationships between individuals with whom they are familiar and to sort conspecific individuals into relational categories. What is perhaps *most* interesting is that they reveal these capabilities in ways that enable human observers to perceive them. Since our own ability to come to conclusions about the behaviour of animals of other species is something we tend to take for granted (and, when we think of it, to think of it usually in terms of methodological problems, the main one of which may be the problem of establishing that an animal is *not* capable of one or another cognitive operation), let us consider some of the evidence that monkeys and apes do make connections between items of social information, sorting that information into categories.

The evidence which is in many ways most impressive comes from research on vervet monkeys and a remarkable series of field experiments conducted by Dorothy Cheney and Robert Seyfarth (1980, 1982a,b, 1985, 1986, 1988; reviewed 1990). Cheney and Seyfarth's basic approach was to record in the field vocalizations emitted by individual monkeys exposed to natural social stimuli (to which the vocalization might be presumed to be a response); then they would play the tape-recorded vocalization through concealed speakers in the presence of other individuals and assess its communicative properties in terms of those individuals' responses (Cheney & Seyfarth 1982a).

Vervet monkeys had provided a paradigmatic example of discrete differences in meaning of primate vocal signals ever since Struhsaker observed that sightings of eagles, snakes, and leopards prompted alarm calls to which companions, even when they had no access to the visual stimulus, responded appropriately – looking up in the first instance, looking down and around in the second, running into trees in the last (Struhsaker 1967; Seyfarth *et al.* 1980). (see Figure 6.1) Cheney and Seyfarth replicated Struhsaker's observations and then went an important step further. They recorded on tape alarm calls stimulated by the appearance of predators and then subsequently noted the responses of monkeys to playback of individual calls in the absence of predators. This enabled them to demonstrate that it was the vocalization that elicited the

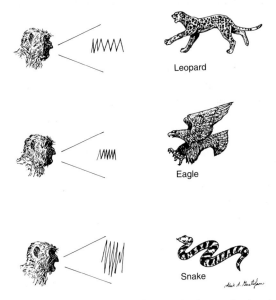

Figure 6.1. Sound spectrograms show differences in alarm calls given by vervet monkeys to leopards, martial eagles, and pythons. In response to leopard alarm calls, monkeys tend to run up trees; in response to eagle alarms, monkeys look up or seek cover (a monkey in a tree may run down out of it and into a bush); in response to snake alarms, monkeys on the ground are likely to stand bipedally and search the ground around them.

response, not concomitant postural-gestural displays and not sight of a predator.

Cheney and Seyfarth were also able to show that varying the length and amplitude of alarm calls had little effect on responses, that the determinative features were acoustic features of the calls themselves. This suggests that the information contained in vervet alarm calls may be relatively independent of arousal state (Seyfarth 1984). The suggestion is perhaps not as convincing as it might be if one could be sure that monkeys responding to the tape had not been able to recognize in a recorded vocalization the identity of the caller. However, and more to the present point, from a receiver's perspective information is information whatever the source, and information which is in some sense of a social character (in this instance the alarm call and not the stimulus that elicited it) may bear a relation to change in the non-social environment, a relation which is congruent with but not identical to the relation it bears to the internal state of the individual uttering it.

Interesting as were Struhsaker's original observations of discretely different warning calls for eagles, leopards, and snakes, these were not as surprising as would have been the case had it been assumed that no functional relationship existed between the appearance of a predator and the concomitant vocalization of vervet monkeys. Predator calls and territory calls are different from ordinary vocalizations, with distinctive acoustic properties relating to signal isolation and transmission across distance (and varying to some degree as these are affected by habitat type). Thus, vocalizations specialized for long distance transmission may differ greatly from one species to another, while vocalizations associated with close, unstressed interaction within a group are at least not obviously specialized in function and appear to vary less across species (Waser 1977, 1982a,b, 1984). Vocalizations of the latter sort hitherto had been regarded as serving primarily phatic functions along with signalling location and, perhaps, individual identity. Cheney and Seyfarth in another set of playback experiments challenged the widespread assumption that information of specific character was conveyed in vocalizations specialized for long range transmission but not in vocalizations associated with the more 'ordinary' routine of foraging and socializing. Utilizing the same record-and-playback technique they had devised for the investigation of alarm calls, they were able to demonstrate that vervet monkeys distinguish at least between vocalizations emitted on contact with another individual of the same group and vocalizations associated with contact across groups (Cheney & Seyfarth 1982b). Monkeys hearing through a concealed speaker playback of a call elicited by appearance of an individual from another group directed their attention in the direction the speaker faced; exposed to playback of a call elicited by the appearance of a familiar group associate they looked toward the speaker, that is, to where the caller would be positioned. Similar experiments revealed that vervets appear to distinguish also, where members of the same group are concerned, vocalizations that signal contact with individuals dominant over the signaller from those that signal contact with subordinate individuals. The semantic implications of these experiments cannot be taken for granted. It may be that a vervet monkey has one response to the stimulus of a familiar individual (or perhaps two, distinguishable by stress constrained acoustic features associated with stimuli by individuals more and less dominant than the signal emitter), and another for individuals he doesn't recognize – this would be a fairly limited categorization system, nor are the responses elicited by playback more revealing. Cheney and Seyfarth appear to acknowledge that the early claims of semantic content were overstated (Cheney & Seyfarth 1988).

Similar techniques were used by Gouzoules and Gouzoules in a study of agonistic vocalizations of the provisioned rhesus monkeys of Cayo Santiago, Puerto Rico, in which they focused on supportive response by kin (Gouzoules et al. 1984). The implications for cooperative behaviour by kin and, more generally, for kin recognition by primates, have been reviewed at length by Sarah Gouzoules (1984). In both these investigations (i.e. by Gouzoules and Gouzoules and by Cheney and Seyfarth), subjects are reported to have responded to signal differences which to human observers are imperceptible or not readily distinguishable but which nevertheless differ in spectrographic form – which is to say that the vocalizations are informative to those for whom the ability to respond appropriately may benefit fitness. In both investigations too it is concluded that the meaning of acoustically similar vocalizations remains constant despite variation in properties such as amplitude and duration, properties arguably more likely to reflect emotional reaction to the evoking stimulus. This fact has bearing on issues of volition and intention, for it suggests voluntary control over the organization and dissemination of social information. Although it is probably as well not to read too many implications for cognitive decision-making, let alone language and meaning, into the voluntary suppression of emotional response, any supposition that the vocalizations of non-human primates are not subject to voluntary control is discredited by such investigations as these. Our interest at the moment is in social information, its character and uses; control, voluntary or not, of any or all components of a vocal signal is clearly important from the standpoint of withholding or conveying information.

Evidence that non-human primates can recognize kinship ties between individuals to whom they themselves are not closely related comes from an experiment by Dasser (1988), who tested the ability of female Java monkeys to recognize mother–daughter pairs from a selection of photographs of individuals known to them. Subject individuals were first trained on a preliminary sample of photographs (all of individuals with whom subjects were familiar) to match photographs of mothers with daughters but not with those of unrelated individuals. They were then presented with a series of photographs of different (but similarly familiar) individuals. Tested thus on a 'first trial only' basis, rhesus monkey subjects proved capable of pairing together daughters and mothers.

That monkeys in matrilineally organized species recognize primary kinship relations between individuals who are not their own kin but with whom they are familiar, and that they respond to individuals partly in terms of those individuals' corporate associations (the same thing, of course, put in other words), is suggested by other observations. Recruitment of aid from relatives in agonistic circumstances is a central feature of primate life. In matrilineally organized groups of savanna baboons, rhesus monkeys, or vervet monkeys it is not uncommon to see spats between individuals expand until family is ranged against family. But vervet monkeys, at least, threaten not only those who threaten them, and the relatives of those who threaten them, they also (in feud fashion) threaten the relatives of those who have threatened their own relatives (Cheney & Seyfarth 1986, cited by Harcourt 1989). According to Harcourt, 'the aim of such indirect aggression is . . . to inhibit future aggression, in the same way that the aim of indirect support is to promote future support' op.cit.,p.233). One might anticipate that post-aggression overtures toward conciliation, as reported in chimpanzees and other species (de Waal 1989a,c), would also be extended along indirect lines, but no such pattern has so far been reported. As interesting from the standpoint of kin recognition are suggestions that in some species of monkeys mating decisions which involve prospective partners in the same or neighbouring groups may be similarly influenced by assessment of genealogical distance (see Quiatt 1988a for review).[1]

Using social information

It is important to note that what is of concern here is not simply the ability to distinguish kinship and other relationships. We know that primates recognize as individuals those animals with whom they are familiar (nor, of course, is such recognition limited strictly to conspecific individuals; primates in captive settings and in nature discriminate readily between individual human caretakers, observers, and others who encroach on their domain on a regular basis.) In this they are perhaps not much different from other species of mammals, birds, and vertebrates in general. That research into individual recognition and categorization of conspecifics in species other than primates is not currently a hot topic may have less to do with the cognitive limitations of, e.g., fishes or amphibians than with problems of methodology, low funding priorities, and (underlying these) relative absence of human curiosity concerning the social cognition of animals whose behaviour in general is so unlike our own. It is interesting from this standpoint that the behaviour of dolphins has recently come under review by anthropologists and primatologists interested in the cognitive bases of social action (Wrangham 1988).

It would be surprising if monkeys and apes did *not* make fine discriminations in social as in non-social domains, if they did *not* integrate those discriminations into patterns, and if, in these emergent relational patterns, they did *not* then discern secondary and tertiary patterns. We have already referred to McGonigle & Chalmers' (1977) demonstration of the ability of monkeys to derive hierarchical structure from consecutive ranking of adjacent pairs, and we have noted the relevance of that experimental test of cognitive activity to social life in a status-ordered society. What is especially interesting about research in social cognition is that it encourages us to ask what it is that primates and other animals do with knowledge about their social world, in the same way that we have long been accustomed to ask what it is that they do with knowledge about the inanimate world and about other species. In short, it is not only the ability of animals to make social discriminations which interests us, it is also the uses to which they, and non-human primates in particular, put that ability.

Whether knowledge in the social domain may differ in essential character from knowledge in non-social domains seems unlikely. What is claimed is 'that natural selection has acted to favor abilities particularly useful in interaction with conspecifics' (Cheney *et al.* 1986). This proposal, the authors argue, can 'be tested only by presenting individuals with logically similar problems':

> Results [from preliminary field experiments] suggest that vervets perform better on tests that use social, rather than nonsocial, stimuli. For example, although vervets appear to recognize with which ranges the members of other vervet groups are associated they seem not to recognize the ranging patterns of other species. When researchers play the calls of species that are habitually found near water, vervets respond similarly regardless of whether the speakers are near waterholes or in arid woodlands.

However, it is by no means clear what interpretation should be given to this consistency of response, regardless of playback context, to the calls of other species. One conclusion of course could be that human researchers and vervets are not in complete agreement as to what constitutes 'logically similar problems'.

3. The importance of social knowledge

One thing is clear: vervet monkeys are very much interested in vervet monkeys. That abiding interest in the character and activities of individuals of one's own species, and more particularly in those with whom one is acquainted on a basis of day-to-day interaction, is what we have in mind when we think of primates of most species as gregarious, and it leads us to a definition of what we must have in mind when we speak of social knowledge. Social knowledge is *knowledge concerning the identity of conspecific others, the character of their behaviour as individuals, and the specific nature of their relationships with one another, including oneself.* In the more gregarious species of primates, ongoing relationships within the local community, group, family unit, etc., provide a layered context for behaviour in which each individual's activity is continually informed and affected by awareness of what others are up to, awareness that is coloured by memory (conscious or no) of a shared past; similarly, his or her every activity is likely to have some informational import, of however small degree, for the behaviour of others. Relationships established through interaction with kin and other long-term associates are of particular importance, mediating an individual's perception of and contact with the non-social world through a long period of dependency, and afterward still conditioning adjustment to habitat and reaction to environmental stress on a day-to-day basis through life. Thus, if social knowledge is not somehow different in character from non-social knowledge, it certainly has a peculiar weight. Not that one or another is more important for individual survival or for the evolution of a species, but as far as the character of certain species' evolution goes it must make a difference that, as Humphrey puts it, for primates like apes and man, whose lives 'depend critically on the possession of wide factual knowledge of practical technique and the nature of habitat . . . [s]uch knowledge can be acquired only in the context of a social community' (Humphrey 1976).

For all animals the surrounding community of conspecific individuals provides a complex set of constraints woven into that matrix in which selection works, encouraging individuals towards behaviour which increases their contribution to the population's gene pool. For highly social creatures relationships within that community, especially but not solely relationships within a changing circle of kin and friendly associates, provide the social tools which individuals deploy in achieving short-term goals and implementing strategies subsidiary to that overriding, more general, and presumably unconscious one of maximizing reproductive fitness. Human evolution has involved considerable development and refinement for corporate use of this social technology, primarily through conventionalization and institutionalization of relationships (see Chapters 9 and 10). However, social technological functions may be discerned in the actions of non-human primates as well, apes certainly, probably monkeys, and very likely the more social of the prosimians, and so research in social cognition assumes new importance.

4. Exploiting social information and managing social relations
However *social information* may be defined, whatever it may consist of
(see our definition of *social knowledge* just above), the kind which is
likely to interest human beings most is that which they can observe in use
(and argue about whether it is being reviewed, integrated, and put to
work consciously or unconsciously) in the interactions of non-human
animals. As an example of what we mean, consider the character of
relationships among free-living chimpanzee females and among females
in the Arnhem Zoo colony. Among free-living chimpanzees of the
Gombe National Park and the Mahale mountains (Tanzania), female
alliance beyond that of mother–daughter pairs is virtually unknown,
while in the Arnhem colony females readily form coalitions 'in defence
against aggression and to influence status struggles among males' (de
Waal 1989a; see also de Waal 1982, 1984.) De Waal, who argues that we
have attended too exclusively to the outcomes of competition between
individuals, ignoring the costs, explains the difference between free-
living and captive chimpanzees in terms of reduced cost of friendship for
the latter. Where food is abundant, as in the Arnhem Zoo, feeding
competition puts less strain on friendly relations.

De Waal's caveat of course is not intended to warn us away from
considering outcomes altogether – this explanation squares nicely with
the socioecological view of social organization as the outcome of inter-
individual competition for local resources of all sorts, including social
resources in the form of other conspecific individuals (Standen & Foley
1989b). Those conspecific others must be considered in their various
guises as, for instance, physical substrates, gamete suppliers, information
transducers, and partners in coalitions (see Chapter 2). What we wish to
note here is the problem faced by observers who may be able to perceive
in coalition behaviour the relationship and some of its apparent functions
(e.g., defence and manipulation of status) but are likely to have less
insight into the informational bases on which manipulative use of the
coalition depends. To use a relationship with another animal as a
behavioural resource must depend on good understanding of its character, and a store of information about the relationship itself. That information store presumably would include data concerning status
differences between partners within the relationship; data of this sort are
frequently, indeed characteristically, displayed in ritual status encounters. However, according to de Waal 'after 6 years of study of the Arnhem
colony – with an estimated 6000 hours of observation – there remain a
dozen female–female pairs in which status encounters have never been
observed.' (1989a:). So, while we may presume in this chapter to speak of

social information and social knowledge as if we knew what we were talking about (and we will), the fact is that it is difficult to characterize information categories with precision because our access to them as human observers is necessarily greatly limited. Does this mean that we should refrain from discussing the uses of social relationships, since we know so little about the informational bases on which use of a relationship must depend? Not at all, as long as we confine discussion to levels of analysis at which we can infer with some degree of confidence instrumental functions from context and sequence as well as form of action.

Social information is drawn from the manifold data bodies on which, along with much else, management of social relationships by competing (and cooperating) individuals depends. It seems likely that for monkeys and apes, as for humans, some but not all of that information may be subject to conscious review (we are not prepared to argue this point to the death, for clearly much depends on how one is to define what is meant by 'conscious'). Our position both with regard to management of social relationships and to artful, intentional use of social knowledge, is in line with some current theories relating to primate behaviour in evolution (see Dunbar 1988; Byrne & Whiten 1988; de Waal 1989a; Mitchell and Thompson in press). However, before we begin discussing relationships as if they were tools put to intentional use (they may be and sometimes are; we are not speaking metaphorically here), it seems appropriate to try to clarify and integrate terminology and concepts which we have been employing with increasing frequency in this chapter.

Self recognition
We referred briefly in the last chapter to the likelihood that monkeys and apes, certainly chimpanzees and orangutans, recognize themselves as individuals. Do they recognize themselves in the same way that we recognize ourselves? First it must be asked, in what sense can it be said that we recognize ourselves? We will not venture into metaphysics but simply note the utility of that narrowly operational definition of self-objectification which has frequently been employed in research with apes and monkeys, i.e., mirror recognition (e.g., Gallup 1977). It seems to human beings no great psychological feat, having effected technological detachment of the visual standpoint, to perceive an imaged self as that entity with whom we are most familiar from the inside, over whose actions and whose *persona* we demonstrably have and can remember having had some control. (It does not follow, of course, that analysis of relations between mind, self and society is no great feat; anthropologists and sociologists owe much to the pioneering work of G. H. Mead, 1934.)

We recognize ourselves, that is, as reflected in mirrors and in photographs, or in a home movie or videotape, carrying the beer from point A to point B until the wife (or husband, parent, daughter, son) catches hold of a sleeve and pulls us off in the direction of C (we aren't the only ones who exercise control over those recognizable entities and images). We can watch ourselves in a store where video equipment is sold, discussing with a salesperson the accuracy of such a representation of the self, the clarity of resolution – and in case of disagreement we can set up additional cameras and monitors, a mirror for control if there is one available, and compare a number of different images, all of which we conceive quite reasonably to be reflections of a single self. What is surprising, on the face of it, is not that we can manage this identification of the self but that other animals, including household pets who appear to recognize ourselves as individuals, should not be able to.

By self-recognition then we mean just this manner of objectifying matter and action, taking a view of our physical selves which is detached both literally, in a technical sense (via mirror or camera), and figuratively. The interesting implications for social behaviour lie of course in the potential for effective manipulation of a self which is privately observed (given continuity of recognition from one instance to another of technical detachment of the observer standpoint) and privately manipulated. Where animals of other species are concerned we can only guess about their ability to objectify their own behaviour in the figurative sense, but a well-founded guess can be made in conjunction with observations of an animal whose viewpoint is similarly detached. Chimpanzees and orangutans evidently recognize their body images in mirrors and video screens and use those images spontaneously for visual self-inspection (Gallup 1977), and to monitor searches by hand for objects within reach but not detectable by unassisted sight (Menzel *et al.*1985). Monkeys and, it has been claimed, gorillas do not appear to recognize themselves in mirrors (Suarez & Gallup 1981; Gallup 1987; Mitchell 1993). However, when comparing and interpreting the meaning of behaviour, we need to bear in mind that similarities and dissimilarities in behaviour observed across species do not in themselves tell us much about similarities and dissimilarities in adaptation, ontogeny, proximate causation, or cognitive function. We need to bear in mind too that negative generalizations are necessarily weak, depending on how we interpret the evidence from single cases. For gorillas, the evidence presented by Francine Patterson for recognition by the gorilla Koko of her mirrored self (Patterson, 1986) should not be passed over. It may be that gorilla self-awareness does not lend itself to, or may actually hinder interaction with, mirror images.

Instrumental use of mirrors and video screens for self-inspection and monitoring body movement nevertheless *seem* like reasonable measures of self-awareness; at any rate they are the best we have at present and are likely to remain so until someone comes up with a measure which is somehow more direct than eliciting responses to visual body imaging. It is hard to think what such a measure might consist of. The situation reminds us of how important visual observation is to scientific research; it is especially difficult to arrive at operational definitions of behaviour which are not grounded finally on visual measures.

Goal-oriented behaviour and purposive action

Suppose we see a gibbon, feeding in the crown of a tree, pause and look about, focus her attention on fruit hanging from the closely adjacent bough of a neighbouring tree, then move immediately out to the terminal branch ends of the limb on which we sighted her, transfer herself carefully over to that other bough, and resume feeding. We might think of the fruit as a visual stimulus and her behaviour subsequent to sighting it as a response to stimulus. Or we might conceive of the fruit as a goal, the transfer from tree to tree an intermediate goal (perhaps only more remarkable than all the other intermediate goals of locomotor progress), and the behaviour we have been observing as organized toward the achievement of those goals. From a behaviourist standpoint, this may be about as far as we should go in analysing her activity. Wanting more information about motivation, internal ('mental') representations, and the larger configuration of stimuli and setting (including her own physiological state), we ought to be satisfied with describing the behaviour we observed as goal-oriented. It is hard to think how one's assessment of it might be improved, still from a behaviourist standpoint, by speculating about underlying purpose and whether or not it would be appropriate to term this goal-oriented sequence of activity an intentional act.

Information review and cost-benefit decisions

As long as we can discuss behaviour in terms of stimuli and the specific responses which follow more or less immediately upon exposure to them there may be little need to refer to whatever decision-making processes may be presumed to underlie behaviour. However, as we saw in the comparison of simple and complex learning (Chapter 5), appropriate responses in a test situation may depend on a configuration of stimuli and features of the context in which they are presented. In tests of complex learning, the interval between presentation of a stimulus set and response by a subject may be gradually or rapidly reduced with learning. What remains is some irreducible constant (subject to variation in repeated tests) which presumably reflects cognitive activity – activity which it is convenient to break into two sorts (at least conceptually; we will not attempt here to distinguish them processually): reviewing information that is relevant to a response, and arriving at a decision between alternative possible responses.[2]

In final analysis, it might seem that the distinction between information review and culminating decision ought to be reducible to precise description of states, either of neural physiology or of some simulated system closely analogous to a neural network. However, work with simulated neural networks suggests that informational concepts such as *storage*, *retrieval*, and *review* may not be applicable at the neural level of behaviour process. Information concepts and symbolic logic simulation models are more useful at the level of whole organism behaviour involving cost-accounted time and energy deployed in strategic action and, especially, goal-oriented social interaction.

Cost/benefit calculation and immediate versus more distant goals
Let us return to the gibbon and to the tree on which we found her feeding, pausing, fixing her attention on some fruit in another tree, moving out 'purposefully' along a limb to its outermost branches, reaching out . . . and this time just failing to grasp any of the terminal branch ends of that neighbouring tree's extended bough. After brief hesitation she returns a few steps along her limb and drops a few meters to an easier crossing; then, safely transferred into the other tree, she makes her way up again to the desired fruit. To suggest that during that brief hesitation she was engaged in cost/benefit calculations might seem speculative, and ponderously speculative at that. We are not quite prepared to move off the behaviourist standpoint, but we will be soon, so let us consider for a moment what it would mean to talk about cost/benefit accounting in this situation. First, costs and benefits must be calculated in terms of some goal. *Immediate* goals call for decision and action based on first order review of behaviour possibilities. End goals that are not also immediate goals can be achieved (if at all) only by achieving intermediate goals in series, each with its own array of choices. Intermediate goals may not present themselves until the time is ripe (and so that final goal, distant but looming, usually is more than and different from the last in a series), but each will call for behaviour appropriate to the current context and hinging, presumably, on some combination of information available in memory's configuration and accessible by direct sensory perception of the external world. For our gibbon, to put it in concrete if greatly

oversimplified terms, a long-range but not so very distant goal is to grasp the fruit she has espied on an adjacent tree. This depends on at least one intermediate and now immediate goal, negotiating a transfer into that tree; how she negotiates it rather obviously must depend on the character of the gap which faces her and on remembered competence at dealing with similar gaps. Routine as this hypothetical example may be (if a hypothetical example may be thought of as routine), it nevertheless will be useful to have in mind for comparison with the uses of social knowledge in social behaviour.

As problems go, the one we have been discussing is routine for a gibbon, but that does not mean that no decision is called for. Let us consider the cognitive situation in Soar terms (see Chapter 4). In Soar terms, crossing a gap is a problem gibbons face tens or hundreds of times daily. It can be accomplished more or less automatically because a good deal of youthful playtime was dedicated to practising skills involved. Inexperience and relatively small size require that young gibbons stick close to central limbs and well-defined, well-connected locomotor substrates. They feed less frequently than do their parents at the terminal ends of the larger horizontal limbs, where no safety net of branches lies extended below (Quiatt, unpublished observations). Individual or social play (for young gibbons lucky enough to have a sibling playmate) may frequently involve repeated scampering circuits over the same looping pathway, high in the central crown of an emergent tree, crossing the same small gaps over and over – and is in this respect much like monkey play, and in fact like the play of most young mammals (Symons 1978). As juveniles increase in size so does the playfield, and gaps traversed grow larger.

Our adult gibbon, looking at fruit in the neighbouring tree, is not faced with a problem of life-shaking proportions. We need not think of her asking herself how badly she really wants that fruit, how much she is willing to risk for it, what her chances are of crossing the first gap alive – doing her arithmetic and solving her problem, to leap or not to leap. In Soar terms, arriving at the gap produces a problem of small dimension, nevertheless one that in certain whole-context circumstances might introduce uncertainty, stimulating review of condition–action rules formulated at previous encounters with problems of similar dimension. For an experienced adult, day-to-day decisions can rarely require more than review of one or two such rules which are immediately relevant and readily accessed. As we have noted (Chapter 4) AI theorists have suggested that human beings probably utilize a few thousand rules in day-to-day life. The exact number of course would vary among individuals, as would the composition of same-sized repertoires. Also, one might expect variability in the character and deployment of individual rules within typological boundaries and across a population. This variability can be assumed to reflect differences in individual history and provides as good an explanation as any of why we cannot predict individual behaviour precisely.

Presumably it would not take our gibbon long to arrive at her decision. She might refer at most to two rules: 1. Cross the gap if it hasn't raised concern, and 2. If the gap raises concern search out a narrower crossing. The second rule takes her immediately out of the problem situation and defines what the next problem situation will involve (contemplation of a narrower gap).

The longest delay times between stimulus and response probably occur in situations to which existing decision–action rules do not appear relevant, either because none have been formulated that *are* relevant or because, while it may be clear that there is a problem, information is insufficient concerning the features by which the relevance of rules at hand might be assessed. In the latter case, the issue cannot be how much time is required for review and evaluation of decision–action rules; it probably boils down to how much time is going to be spent delaying action, dithering, until a decision is arrived at to get on with other business. The former case is more interesting, as it seems to be what thinking is all about – some process of comparing and evaluating rules which do not quite cover the present situation but come closer than do others, a broader review of information concerning behaviour settings not closely related to the present, vicarious trial and error, and, eventually, solution of the problem and formulation of a new decision–action rule. The new rule, if it proves applicable to similar situations will ease future decisions, *that* problem at least having been eliminated (Laird et al. 1986). It may not prove broadly applicable, in which case a more useful rule remains to be formulated. Of course, a rule may be accepted that is not as good as it might be and, worse, may fail subsequently to be rejected in contexts in which it is even less acceptable. Decision-makers may in this fashion be led up the garden path, which is exactly the way things frequently happen in life, and in evolution too.

Deception and intention

References to *condition–action* rules in the last section carry for us stronger connotations of purpose than does the phrase 'goal-oriented behaviour', designed after all to be neutral with respect to just those connotations. In considering most behaviour, including that of our hypothetical gibbon, whom we can now leave feeding comfortably again, there is nothing to be gained by introducing ideas of conscious intention, purposeful action, and the kind of reflective thinking that we ordinarily associate with the making of human decisions (the 'big' decisions, that is, and keeping in mind that for an individual a big decision is not whether to declare war but what to wear to the dance). Social behaviour appears to be another matter (what to wear to the dance is of course a social decision, involving sometimes complex consideration of what reactions and interactions the right choice – so one speculates – might produce.) Here the case of deception is suggestive; for many, deception as practised by chimpanzees provides convincing evidence that among non-human primates chimpanzees at least are capable of formulating long-range strategies and initiating whole series of short-term tactical sequences, which are strategy-derived but contingent in detail on the immediate contexts of social behaviour (e.g., de Waal 1986; Whiten and Byrne 1988b). Old World monkeys, some would argue, are capable of the same. Such behaviour might reasonably be described in terms of purposive *social action*.

Deception in non-human primates ranges, arguably, from withholding information to actively misrepresenting events and intentions. We say 'arguably' because several caveats have to be entered. First, not everyone would agree that withholding or dissembling information and misrepresenting one's intentions should be lumped together in a unitary concept. Second, and related to that first reservation, is the fact that it is a good deal easier for a human observer to recognize when an animal is (or seems to be) withholding information than when he is (or seems to be) misrepresenting what it is he has in mind to do. We agree with those who would leave interspecific mimicry and environmental camouflage out of account, for reasons which we will give in a moment. Third, and this relates to those parenthetical 'seems-to-bes', is the fact that, under ordinary circumstances, animals subject to frequent deception – animals as intelligent as monkeys and apes anyway – soon learn not to be so easily fooled (Quiatt 1984, 1988b). And so deception ought in theory to occur infrequently, as in fact it does (at least recorded observations are relatively sparse and anecdotal, which unfortunately makes for difficulties in accumulating a reliable body of data on comparable events – but see Whiten & Byrne 1988b). This should not be a particular problem; there are, after all, other behaviours which are problematic and infrequent. However, deception is perceived by many as central to the interpretation of intention (e.g., Mitchell & Thompson 1986), and caution is advisable where it comes to constructing what amounts to a theory of primate cognition on the basis of anecdotal observations of behaviour (Kummer *et al*.1990, and see Chapter 7).

We can compare withholding information with misrepresenting intention, and at the same time illustrate in particular detail the conceptual problems just noted, by considering two widely cited instances of deception, one at either end of the scale of increasingly complex action. But first let us say why we leave camouflage and interspecific mimicry out of account. We do it for two reasons, and they are rather obviously related. First, the stimuli which evoke what turn out to be inappropriate responses on the part of predator or prey (or, in the case of camouflage, the dissimilation of stimuli which might otherwise evoke appropriate responses) are not learned but are 'fixed' characters of a species' physical morphology and/or communicative behaviour – e.g., the sticklike body structure of a mantis (Wickler 1968) or the interval between flashes of light emitted by a male firefly that lures to death males of another species attracted to what they take to be a conspecific female (Lloyd 1986). Such characters are the products of highly particular histories of selection and adaptation. Second, as we have noted, and as these examples suggest, the deception in cases of camouflage and mimicry is interspecific. This general observation is tested by a few instances of sexual 'mimicry' within species (Wickler 1967, 1968), which are dubious in the sense that it is only by a stretch of the imagination that one can regard responses to stimuli in most such cases as inappropriate (but see Trivers 1985). As dubious as any are the two or three purported instances of sexual mimicry in primate species, e.g., by gelada males (Wickler 1968) and human females (Morris 1967). In short, mimicry involves signals more or less constant in character directed across species or, in a very few cases, across classes of individuals within species. Indiscriminate application of communication theory to analysis of signals transmitted across species and signals transmitted within species can obscure important differences between the two; the main differences should be clear from the foregoing discussion. We are interested in deception which involves social interaction within species and signals directed toward individuals in contexts of familiar interactive behaviour, as in the catalogue of 'tactical deception' compiled by Whiten & Byrne (1988b) and presented in Box 6.1. It probably is significant that the great majority of reported instances of intraspecific deception involve interactions within groups, that is between individuals likely to be familiar with one another and to have pre-established expectations of behaviour based on that familiarity.

In Chapter 7 we discuss the intentionalist approach (to analysing social behaviour) at some length, and we present there several examples of

Box 6.1. Types of tactical deception. (From Whiten & Byrne 1988b.)

Whiten and Byrne have sorted records of tactical deception into five major functional classes, divided in turn into 13 subclasses. The *agent* is the individual performing the deceptive act and the *target* is the individual who poses the problem which the *agent's* behaviour deals with.

1. Concealment

The *agent's* behaviour functions to conceal something from the *target*.

Hiding from view. The *agent* hides an object, or a part or whole of itself, by screening it from the *target's* view.
Acoustic concealment. The *agent* acts quietly, such that the *target's* attention is not attracted.
Inhibition of attending. The *agent* avoids looking at a desirable object when such looking would lead one or more *targets* to notice it.

2. Distraction

The *agent's* behaviour functions to distract the *target's* attention away from some locus at which it is directed, to a second locus.

Distract by looking away. The *agent* distracts the *target's* attention by looking away at another locus in such a way that the *target* also looks there.
Distract by looking and vocalization. The *agent* distracts the *target's* attention by looking away at another locus and vocalizing in such a way that the *target* looks there or at least loses the original focus of attention.
Distract by leading away. The *agent* leads the *target* away from the first locus to another one, allowing the agent to return to the first location free of competition.
Distract with intimate behaviour. The *agent* shifts the *target's* attention to some part or extension of its own body, which is highlighted posturally or gesturally.

Box 6.1. *Continued*

3. Creating an image

The *agent's* behaviour functions to create an impression which, rather than merely affecting the *target's* attention as above, causes the *target* to misinterpret the behaviour's significance for itself in other ways.

Present neutral image. The image is non-threatening just in the sense that it is of little or no significance to the *target*.
Present affiliative image. The image is not merely neutral, but is affiliative.

4. Manipulation of target using social tool

The *agent* manipulates one individual, the *tool*, so as to affect the *target* to the *agent's* advantage.

Deceive tool about agent's involvement with target. The *agent's* behaviour misleads the *tool* about the significance of the involvement of the *agent* with the *target*.
Deceive tool-1 about tool-2's involvement with target. The *agent's* behaviour misleads one *tool* about the significance of the involvement of a second *tool* with the *target*.
Deceive target about agent's involvement with tool. The *target* is deceived about the significance of the results of the *agent's* action on the *tool*.

5. Deflection of target to fall-guy

The function of this behaviour is to divert the *target* who poses a problem towards a passive victim, the *fall-guy*.

deception. Whether they document satisfactorily the operation of some distinctive process we can characterize as conscious intention is not as important as is the fact that in some instances behavioural goals appear to be realized as a consequence of one individual's misapprehension of where the behaviour of another is tending, mistaking an apparent goal of that behaviour for the real goal, a misapprehension which the behaviour of the other seems designed to encourage. Whether or not animals act with or without conscious purposes, that is, whether or not they act with or without reflection on the purpose and character of their actions, and on possible interpretations of that purpose and character by associates, is hard to say; too protracted speculation about the essential character of social behaviour can lead to such logic-chopping exercises as wondering whether deception is not most successful when the deceiver himself is deceived, and, in that case, whether self- recognition may be a help or a hindrance to deception (e.g., Lockard & Paulhus 1988; Trivers 1985). These are interesting questions, to be sure, and perhaps the issue to be addressed is not whether they can be answered but at what levels of analysis answers may be sought and for what purposes they may be required. Concepts may be very useful at one level of analysis without being useful at all at other levels.

A useful starting point is to say that animals behave *as if* their behaviour were premeditated. This may, however, encourage us toward anthropomorphic analyses of social behaviour. Analyses of deception commonly involve interpreting and predicting social behaviour by attributing to animal actors goals and motives which are similar to those which we know influence our own human behaviour. It is especially tempting to attribute something like human motives to chimpanzees, whose intelligence seems in many ways so like our own. Of course, the apparent continuity may lie not so much in the specific character of intelligence applied to or evolved in a social domain of activities as in the general similarity of social behaviour across ape and hominid species. Nevertheless, the evidence for objective self-recognition in chimpanzees is convincing, and chimpanzees not only engage in complex, multi-stage sequences of behaviour involving the manipulation of objects, other chimpanzees, or both (see, e.g., de Waal 1982 and Savage-Rumbaugh 1986), they do so with every appearance of monitoring and regulating their own behaviour and that of associates (see pp. 196–99, this volume for discussion of communication and cooperation between Austin and Sherman, two of Savage-Rumbaugh's chimpanzee subjects). It seems, then, not unreasonable for human observers to assume, in the case of a chimpanzee apparently engaged in deceiving another chimpanzee, that the deceiver is objectively manipulating a particular representation of himself and his intentions, monitoring his associates' responses and regulating his behaviour accordingly, in the same way that he would if he were grooming a companion or manipulating a physical instrument. And the instances of tactical deception which Whiten and Byrne have collected (1986, 1988b) suggest that Old World monkeys are not far behind apes in this respect. Deception therefore is of interest to many primatologists because of the light it seems to shed on the objective, instrumental projection of the self in action, allowing us to think of monkey and ape social behaviour *in terms of* intentional action.

Putting aside for now the thorny issue of intention, we can say that deception is little different from other forms of exploitative behaviour. Let us consider then the ways in which individual primates manage their interactions and, over time, their relationships with others. They may of course manipulate others in ways which we have discussed, by withholding information or conveying false information. But they may also manipulate others by sharing information that is correct, by, for instance, emitting predator alarms which stimulate appropriate life-preserving responses by companions (Struhsaker 1967; Seyfarth et al. 1980a,b). And, finally, they may exchange information and provide mutual assistance in short-term and long-term friendships with particular individuals (Smuts 1985; Strum 1987). These of course are categories in a logical continuum from active exploitation of one party by another to equally active mutual cooperation. It would certainly be in error to suppose that, for instance, in a group of rhesus monkeys, distantly related matrilines were more or less steadily withholding information from one another, closely related matrilines were occasionally sharing, and within matrilines individuals were ever engaged in a harmonious round of mutual backscratching. There is in fact, however, at least a tendency for things to work out somewhat like this, since, within groups, close associates are most likely to share information and hence least likely, given that fund of shared information, to succeed in deceiving one another.

There are close and obvious connections between deception and information sharing. Whiten and Byrne have collected few or no reports of inter group deception. This may be partly an artifact of sampling problems; deception occurs infrequently, and, for a variety of practical and theoretical reasons, students of primate behaviour have tended until recently to focus their attention on social interaction *within* groups. However, the reason that successful attempts at deception are few and far between must have to do with the ability of regular and probably even occasional associates to 'see through' would-be deceivers with whom they are accustomed to share information, and so one would expect deception to be practised (and practised more successfully) at greater relative frequency *across* groups. That it apparently occurs at lesser, not greater relative frequency, suggests that sharing of perspective at some optimum level is essential to deception (Quiatt 1988b).

Vasek (1986) has discussed 'perspective-taking' in human children, in connection with the development of deception, and the importance of shared perspective to successful deception is also illustrated in human behaviour by the reluctance of professional footballers to engage in games with rank amateurs. According to Mawby & Mitchell (1986), professional footballers are concerned about injuring *themselves*, not, as one might suppose, about doing injury to less-skilled amateurs. It is difficult, apparently, to fake out an opponent who has only a dim understanding of the moves which are being put on him. At another level of human experience, common or translatable language ensures widespread cultural commonalities, making it feasible in theory to deceive others on a species-wide basis (Quiatt 1988b), something which seems absurd even to contemplate where other primate species are concerned.

No assumptions about a single origin for language are buried in that observation. However, if current languages are rooted in the genesis of anatomically modern *Homo sapiens*, and not in prior divergences, one cannot but wonder – in view of the importance of common language to shared perspective, and of shared perspective to deception – whether and in what ways inter-group deception could have been deployed by late middle pleistocene hominids in a species undergoing (as some have argued, e.g., Stringer & Andrews 1988) divergence.

5. Conclusion

But, to return to our consideration of mutual exploitation of individuals within groups, even within a close-knit rhesus monkey matriline there are frequent disharmonious interruptions in the exchange of information, bickering between cousins, tiffs between sisters, and rebellions of daughter against mother. Interestingly, quarrels in these familial circles are almost always settled internally (Quiatt 1966; Hughes 1988; Datta 1989) despite – or perhaps at these levels of familiar matrilineal interaction it were better to say in consequence of – the tendency for participants in an agonistic bout to enlist support from friends and relatives. Mothers can usually be relied on to settle a spat between siblings who can't settle it for themselves, or an older sibling may do the job. Should cousins involve themselves in a sibling-sibling quarrel, or if the quarrel is between young cousins to begin with, escalation is still not likely to occur beyond the point at which the female head of one or the other of the sister 'households' involved decides to take a hand. *Her* sister of course may not approve of the means she chooses to apply, and if that disapproval is forcefully expressed then male friends of either party may introduce themselves and use the occasion to pursue their own personal programs for advancing reproductive fitness. In short, quarrels within the wider matriline or across matrilines involve correspondingly greater numbers of allies and, for participants, second order social calculations and redirection of aggression along lines consonant with original targets (Harcourt 1989). However, disagreements between close relatives and good friends ordinarily do not escalate so far. They are in fact most likely to be terminated by the initiating parties themselves, one of whom sooner or later – and sooner usually, once active aggression has subsided, than later – solicits or proffers grooming. Similarly, in the less frequent but more serious altercations between matrilines, the more individuals involved on either side, the more extensive the conciliatory grooming afterward (de Waal 1989a).

Quarrels do not always end in harmonious resumption of the status quo. Chimpanzees engage in conciliatory behaviour, but they are also capable of exacting revenge. And, as is evident from cases of gradual reversal of status accompanied by repeated challenges of one male by another with series of aggressive incidents extended over days and weeks, aggressive encounters in any event are not isolated pieces of activity that begin with a threat and end with a conclusive punishing bite or a conciliatory overture (de Waal 1982). Certainly a recorded bout of overt aggression usually closes with some decisive action, but that is at least partly a matter of the observer drawing a useful categorical boundary; that we draw such boundaries should not fool us into thinking that aggression is somehow resolved either by stern administration of justice or by arbitration and a lasting truce. In the past, functionalist and group selectionist assumptions concerning role responsibilities of high dominant males certainly tended to support such a view – i.e., when 'settling fights' was placed high on the list of role functions prescribed for 'leader' (*vide* alpha) males. But under normal circumstances aggression either ends in eventual if temporary reconciliation of parties involved, without apparent change of status of participants, or it concludes with changed status *and* reconciliation, if temporary, of parties involved (de Waal 1989c; cf. the notorious London Zoo 'Monkey Hill' colony of Hamadryas baboons as portrayed by Zuckerman 1932). Thus, while we need not conceive of primate social life as consisting of endless cycles of conflict and reconciliation, the idea that groups are composed of interacting individuals engaged in continual exploitation of one another should not conjure up onesided images of primate life as either enviably harmonious or appallingly deceitful and manipulative.

Notes

1. It has been argued by Grafen (1990, 1991a) that the term 'kin recognition' should be restricted in use to 'recognition by genetic similarity detection' (1991a:1091). The underlying issue, says Grafen, is whether 'animals really

recognize kin in a way that is different from the way they recognize mates, neighbours, and other organisms and objects' (Grafen 1991b:1095). Our discussion of kin recognition is confined to primates, who (with obvious qualifications relating to the use of language by human beings) can be said to recognize kin in very much the same way that they recognize nestmates, siblings, and some close neighbours – that is, on the basis of developmental familiarity (Bekoff 1981; Byers & Bekoff 1986, 1991; Quiatt 1988a). We agree with Grafen that terminological distinctions are important but not that using 'kin recognition' in this looser sense is necessarily 'misleading, presumptive and unhelpful'. We use it because it seems to us meaningful and helpful in comparing human and non-human patterns of association and ways of categorizing associates. Human beings' recognition of who they are related to is reinforced by and in many instances is primarily determined by developmental associations. It would seem to us at least as misleading, presumptive, and unhelpful to seem to suggest, through a too careful use of substitutive terms, that human beings therefore do not recognize kin.

2. But isn't this just a rhetorical device to 'reify' decision-making? Is a decision anything more than the end point of information review? Can't we assume that information review will result in action one way or another? No, we think not; these are behaviourist devil's advocate questions. For human beings, at least, information review may result in a decision and no action – or in behaviour which we are usually quicker to identify and give a name to when we see others doing it than when we are doing it ourselves, i.e., dithering. This does not, however, dispose of the problem discussed in the next paragraph of the text.

7 *Intentionalist interpretations of behaviour*

1. Introduction

Discussions of animal thinking (Griffin 1976, 1978, 1984; Hoage & Goldman 1986), animal deception (Mitchell & Thompson 1986; Byrne & Whiten 1988, Whiten & Byrne 1986, 1988a,b), and primate social cognition (Humphrey 1976, Smuts *et al.* 1987; Byrne & Whiten 1988) are certain to interest us as human beings, especially in their intentional aspects, because to the extent that non-human animals engage in planning, role playing, deliberate cooperation, and careful manipulation of individuals, relationships, and alliances, their behaviour must involve features which are not dissimilar to those which characterize human action as defined by Reynolds (V. Reynolds 1976). And, because it seems self-evident that there must be close connections between culture and cognition in human evolution, self-reflexive intentional behaviour in other species is of particular interest to anthropologists.

2. Reflexive thinking and intentional action

The case for self-reflexive strategic thinking and intentional action is difficult to establish on the basis of available evidence, and prolonged definitional debate could ground the enterprise at the start. The most parsimonious view of intentional behaviour is that it is synonymous with goal-directed activity, and if that were all there were to it few would deny that non-human primates and indeed all animals act intentionally. Action controlled by a reflexive, self-conscious intelligence is quite another matter (Harré 1984). For one thing, put in just those terms it sounds a little too much like the Uebermensch, if not Nietzsche's Zarathustra then at least a thinker along the lines of Shaw's Superman or maybe Shaw himself, brisk, intelligent, and strangely juiceless. It is hard enough to think of our own human behaviour in terms of such cold-blooded computation, much less that of non-human animals, willing though one might be to grant them non-reflective life as Cartesian engines.

However, all that we have in mind when we refer to reflexive,

165

self-conscious intelligence determining behaviour is whatever cognitive processes become engaged when, for instance, a human being contemplates alternative social actions and tries to decide which to pursue, wondering whether one or another might be modified to make it more effective, wondering perhaps too, in that connection, whether his other aims will be apparent, and possibly reflecting that on the whole it might be better if they were not. Admittedly, all this is a great deal to *have* in mind, but it seems a fair representation of self-reflexive social cognition (what one of us, Reynolds 1986, has called 'social thinking').

To explore the question of animal awareness from a comparative evolutionary standpoint, with human beings the final referent species, makes sense to us as evolutionists, as anthropologists and, inevitably, as human beings, for human beings cannot hope finally to separate the question of animal thinking from the question whether other animals think as human beings do. We intend, however, for reasons that we trust most readers will agree are good and sufficient, to beg the final questions: *Are* other animals capable of reflexive self-awareness? *Is* the behaviour of other animals intentional? *Do* other animals think as we do? First, and apart from the fact that there are no unequivocal answers to questions about the presumed mental character of other species, to deal with such questions at the philosophical level would return us immediately to the definitional difficulties noted in the paragraph preceding this, issues which, however interesting and important, are a sargasso sea to any voyage of discovery. For the fearless, who would venture into that sea, there has been a swell of literature (see, e.g., Griffin 1976, 1982, 1984; Harré & Reynolds 1984; and various contributions to *Behavioral and Brain Sciences* beginning with Dennett 1983). Second (and, for some, good reason to avoid debate on the philosophical level), such questions probably can be handled better at the level of biological nuts and bolts, in detailed comparison of the perceptual/cognitive anatomy, electrophysiological process and actual neural system features on which differences between species depend. An overview of a few such differences, essential to comparison at the level of social interaction, which is our primary concern, has been provided in Chapter 5. Third, and most persuasive as far as present goals are concerned, is the unlikelihood that philosophical discussion of the biological limitations of individual organisms, particularly of hypothetical threshold or *sine qua non* features of perception or cognition deemed essential to intentional action at increasing orders of complexity, can contribute much to understanding how the evolution of social cognition, of social life, and of culture has proceeded.

What seems to us more important, if we are to acquire that understanding, is discovering how social information, having been acquired by individuals in groups, is disseminated, stored, retrieved, and processed (added to or erased, transformed, returned to memory) by means and in contexts which are afforded only in group life.

Before we turn to consideration of the uses and character of social information in evolution, and partly as a final explanation of why it is better not to dwell on does (s)he/doesn't (s)he questions of intention and reflective self-awareness, but to engage in informational analysis (to turn away in that particular direction), we want in this chapter to take a closer look than we have in previous chapters at problems in the way of the first, namely problems of presenting and interpreting evidence for intention.

3. Deception and intention

The best evidence for intentional action in non-human primates comes from anecdotes of deception, accounts of cooperation, and descriptions of instrumental manipulation (of objects and in tool manufacture), in all of which observers can reasonably expect to discern gaps between goals that an animal is conceived to have had in mind (clearly interpretation of the facts is as important as collecting and presenting them) and achievement in practice. Object manipulation and tool behaviour however provide only very indirect evidence of intentional action *in conjunction with social cognition*, and cooperation has received far less attention than has deception from researchers interested in examining intentionalist interpretations of behaviour (Mitchell & Thompson 1986; Quiatt 1984; Byrne & Whiten 1985, 1988; E.O. Smith 1987; Whiten & Byrne 1988b). We will therefore focus our discussion on deception.

Problems with anecdotal evidence

A basic difficulty with the evidence for deception in nonhuman primates or in other animals is that it is inherently anecdotal. While it would be nice to have repeated series of instances in which animal A practices deception on animal B, what might be termed the 'cry wolf' factor insures against collecting these. Any series of consistently successful deceptions practised by an animal on regular companions invites the question as to whether those companions are really being deceived, while deception by non-human primates of individuals that are not regular companions is hardly likely to (indeed, cannot by definition) occur on more than an infrequent basis. A subadult male (aged 4 years) in a group of rhesus monkeys observed by Quiatt (1966, 1984) on Cayo Santiago, Puerto Rico, was consistently successful in cajoling three younger playfellows and a timid adult female (aged 7 years) into submitting to grooming, only to surprise the groomee sooner or later on almost every occasion with a sudden nip of the teeth or rough handling as soon as he or she appeared to have relaxed. Here it is a nice question whether victims quite obviously wary of their companion's behaviour were deceived anew by each invitation to submit to grooming or, considering the context of this group's social life (ten animals in a 100' × 100' enclosure, neither the group nor the enclosure unnaturally small by captive standards, nevertheless a restrictive social setting with no escape), whether target animals might not have been trained into a response pattern of 'learned helplessness' (Maier & Seligman 1976; Maier & Jackson 1979). In that case the surprise would seem to consist primarily in the timing of the attack, nicely adjusted to a gradual, inevitable accommodation of a groomee's wariness and initial resistance to the physical satisfaction of being groomed. We choose the latter consideration, and so we prefer to describe the case in terms of cajolery and submission, not deception, but it is worth remembering that for primates including ourselves there is no escape anywhere from society, hence no escape finally from deception. In the hermit's society of one there remains self-deception, which may be supposed to flourish there at least no less abundantly than elsewhere. This is not to suggest that there is larceny in every heart or that victims are necessarily accomplices in the crime – victims of confidence games (Maurer 1940, and see Mamet 1987) are at least more obviously engaged in performing victim *roles* than are the targets of muggings – only that there is greater definitional overlap between concepts like *deception, learned helplessness,* and for that matter *cooperation* than we are likely most of the time to suppose.

Of necessity, then, reports of deception tend to be anecdotal in character (see Whiten & Byrne 1986, 1988b). Let us take a close look at two anecdotes of deception which have frequently been cited as evidence for intentional behaviour. The first is deceptively simple, apparently hinging on concealment of information, no more. The other is by comparison complex. Both, however, illustrate how extraordinarily difficult it may be even for the most astute observer of primate behaviour, not to construct a case for *intentional deception*, but just to summarize the evidence objectively and to organize interpretation so as to present a case for deception which is acceptable on its own merit. (Deception of course can occur without intention just as virtue can be maintained in the absence of opportunity, and is just as interesting.)

The case of Figan and Goliath

The first anecdote involves Figan, a Gombe Stream chimpanzee well-known to those familiar with Jane Goodall's descriptions of chimpanzee life. It is by no means beside the point that Figan is a character with a personality strikingly his own, in a retinue of chimpanzee characters whose history has been related in narratives designed for a broad readership (Goodall 1967, 1971); nor is it beside the point that the author of those narratives enjoys wide repute as a field primatologist. The anthropomorphic terms and intentionalist phrasings in which anecdotes of deceptive behaviour tend to be couched, necessarily perhaps, but awkwardly, are less jarring when they involve a character with whose personality we are familiar and are presented by a scientist whose authority is well established. Figan of course is not a character in the conventional sense of a figure in a play or novel. Nevertheless, he *is* an idiosyncratic character in the literature of primate behaviour, described by an author whose integrity we trust and whose close acquaintance with Figan predisposes us to accept her representation of this individual's actions as authentic, her interpretation of those actions as valid. We must therefore take care to distinguish the character from the animal whose actual behaviour has afforded its construction. Since most of us will not have been able to observe the animal behind the construction, it goes without saying that distinguishing between the two will be difficult when we are presented with representations and interpretations of Figan's behaviour.

When provisioning was initiated at the Gombe Stream Reserve on a regular basis, adult males tended to monopolize the concrete boxes in which food was distributed. 'And so,' as Goodall tells the story in her popular book, *In the Shadow of Man*:

> . . . we took to hiding some fruits, one here, one there, up in the trees; youngsters such as Figan quickly learned to search for these whilst the adult males were busy loading up at the boxes. *One day, some time after the group had been fed, Figan suddenly spotted a banana that had been overlooked – but Goliath was resting directly underneath it. After no more than a quick glance from the fruit to Goliath, Figan moved away and sat on the other side of the tent so that he could no longer see the fruit. Fifteen minutes later, when Goliath got up and left, Figan, without a moment's hesitation, went over and collected the banana.* Quite obviously he had sized up the whole situation: if he had climbed for the fruit earlier Goliath, almost certainly, would have snatched it away. If he had remained close to the banana he would probably have looked at it from time to time: chimps are very quick to notice and interpret the eye movements of their fellows, and Goliath would possibly, therefore,

have seen the fruit himself. And so Figan had not only refrained from instantly gratifying his desire, but had also gone away so that he could not give the game away by looking at the banana. Hugo and I were, properly, impressed. (1971:103–104, emphasis added)

The first thing to say about this episode is that we (i.e., the present authors) accept Goodall's account as a true and accurate rendering of something that actually transpired. We do not question either the representation of events (as set off in italics) or the interpretation that follows. To this profession of belief, however, we need to add the qualifier that belief is what is involved, for there seems no way to demonstrate the truth of Goodall's assertion that Figan sized up the situation and acted accordingly; rather, the account in its entirety is *authenticated* for us, as presumably for most readers, by our trust in Goodall's experience at observing and recording chimpanzee behaviour and in her accumulated understanding of this particular animal. None of which makes her account more than an anecdote, but certainly it is a far cry from the anecdotes one might anticipate hearing on a radio 'talk' show, with the telephone topic for tonight: 'How smart is your pet?' *Any* anecdote of course can be subjected to criticism on the basis of textual analysis, its terminology (intentional throughout, in this instance) called into question and alternative interpretations offered at various levels of analysis, e.g.: Figan did not size up the situation, he was simply 'inhibited' by Goliath's presence; a banana in a tree with Goliath underneath it is not the same thing as just a banana in a tree; Figan went away and forgot about the banana but remembered it again on seeing Goliath depart; etc., etc.

This kind of question-raising can be useful, where there is some point to it beyond enumerating endless possible interpretations of behaviour, each as real and as unreal as the next. However, as we have already remarked, the 'truth' of this account, either as an accurate description or as an adequate explanation of natural phenomena, cannot very well be demonstrated; we must put our confidence in its authenticity, in effect as a 'true story' of what happened one day when Jane Goodall was observing chimpanzees. Goodall may in fact have been observing Figan in particular and, using a focal animal approach, recording his activities in detail. However that may be, all that this account contains are details relevant to the story.

Thus, what we put our confidence in is not just Goodall's interpretation of the facts but her presentation of them, and we are on shakier ground here than in the case of problem-oriented descriptions of behaviour, in which an observer has a pretty good idea to begin with what he or she is looking for, sufficient at least to dictate the kinds of data to be collected. Such data collection has its own problems, of which bias of expectation is inevitably one, but guarding against that is at least easier than eliminating bias from *post hoc* recollection. The main problem with anecdotal accounts of behaviour is that the meaning of events on which the anecdote subsequently is to depend tends to clarify itself only gradually or to seize the observer's mind with a concretizing Aha! somewhere toward the conclusion of a sequence that was not anticipated and cannot be replayed. Where this anecdote is concerned, a reader's grant of authenticity depends on confidence that Goodall's experience has enabled her, like Figan, to 'size . . . up the whole situation' *at the start*, and, having sized it up correctly, to seize on the essential detail of Figan's behaviour as events progress over a 15-minute period – not the easiest task in the world, given that every one of Figan's behaviours in between discovering the banana and claiming it at the end must have been, with the exception of his one quick glance at Goliath, of negative character.

The case of Nikkie and Spin

Consider now an anecdote in which the series of events critical to interpretation of intention controlling action appear to have occurred in a very short span of time, perhaps 4–5 seconds at most. The account which follows, and the figure to which it refers (Figure 7.1), are taken from Frans de Waal's categorical analysis of deception practised by chimpanzees at Arnhem Zoo (de Waal 1986). At Arnhem Zoo the chimpanzee Nikkie has acquired a reputation 'both among people and apes . . . for his dangerous habit of throwing well-aimed, heavy objects'. In charging displays Nikkie likes to carry a weapon, which he often conceals behind his back. On this occasion, he is in aggressive pursuit of 'a female, Spin, who sought cover behind a tree trunk'.

De Waal's interpretation of the action depicted in Figure 7.1 runs as follows:

> Nikkie started to turn to the left, and Spin responded by moving to the right. At the moment Spin appeared around the corner [*sic*], Nikkie threw a brick, but almost without losing speed, so that he was able to catch his victim when she jumped back to the left in order to avoid the projectile.
> If Nikkie had not had a weapon with him, he almost certainly would not have continued to move to the left once Spin appeared at the other side, because this would have given her an easy escape. That Nikkie anticipated her jump into his arms was evident from the fact that he did

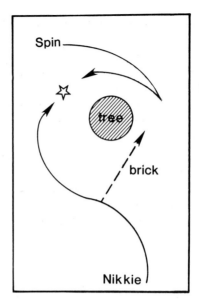

Figure 7.1.   Schematic illustration of an episode at the Arnhem Zoo, in which an adult male chimpanzee, Nikkie, pursued a female, Spin, who sought cover behind a tree trunk. Nikkie started to turn to the left, and Spin responded by moving to the right. At the moment Spin appeared around the corner, Nikkie threw a brick, but almost without losing speed, so that he was able to catch his victim when she jumped back to the left in order to avoid the projectile. The smoothness and speed with which the whole manoeuvre was executed suggest to de Waal that Nikkie may have planned as many as five steps ahead: 1. I move to the left, 2. target will move to the right, 3. I throw stone, 4. target will jump back, and 5. I reach target. (de Waal 1986:235.)

not wait to see her reaction to the projectile. The smoothness and speed with which the whole manoeuvre was executed even suggests that Nikkie may have planned five steps ahead. The five hypothetical steps are: (1) I move to the left, (2) Target will move to the right, (3) I throw stone, (4) Target will jump back, and (5) I reach target. (p. 235).

'If we accept this explanation of the sequence,' de Waal suggests (p. 235), 'it would represent a unique case of deception, i.e., not involving false intentions or signals, but the creation of a false escape opportunity.' But if we accept the explanation, in which again the account foreshadows interpretation and the intentional language of interpretation bleeds back into the description of events, it is again because the story, in this case fast-moving, has a stamp of authenticity derived from its being recounted by a well-known primatologist with established expertise and thorough familiarity with the animals whose behaviour is being described (de Waal

1982), not to mention a style of writing which is as engaging as that of Goodall.

4. The importance of documenting intentional behaviour

Perhaps we should remind readers again that we do not question the authenticity of the accounts to which we refer in this chapter and note as well that authenticity is not an issue peculiar to descriptions of primate behaviour: the characteristically 'objective' constructions of most scientific writing function in part to establish the authenticity of the scientist in his role as detached observer. In *small* part, objectors might protest, and we would agree that such a function is best explored as an aspect of style or rhetoric and in linguistic terms – as, for instance, in textual analyses by literary minded ethnologists such as Geertz (e.g., 1983) and contributors to Clifford & Marcus (1986); it seems to us by no means to warrant deconstructionist revision of the history of science (cf. Latour 1987 and Haraway 1978, 1989). We are concerned, not with textual deconstruction, but with emphasizing how difficult it may be for even the most practised, most responsible, and most articulate observers of primates in nature and in captivity to establish a reliable evidenciary base for assertions concerning the revelation of intention embodied in action. In the next chapter we will suggest that an informational analysis of social interaction – that is, an analysis focused on the different kinds of information available to animals and on formal differences in processes governing storage, retrieval, and integration in the solution of everyday problems – may be more productive than intentional analysis.

If, in the story of Figan, we have a kind of interior drama in which next to nothing happens – Figan sees a banana which he cannot retrieve in the presence of Goliath, removes himself in order not to betray to Goliath what he has seen, and later returns to claim the banana – the story of Nikkie and Spin is packed to the hilt with action. It is, really, nothing but action, the meaning lies in the action, and since all the action transpires in the course of a moment or two, what is there to say about it that de Waal has left unsaid? Well, it would be helpful at least to have a more exact idea of distances involved than can be gained from the diagram of Figure7.1: an accurately hurled missile at 6 or 8 feet from target is not the same as one hurled from 15 or 20 feet. Again, one could wish for a clearer image of Nikkie hurling that brick 'almost without losing speed', raising his arm, presumably more orthograde in posture, with upper torso oriented for that one split second more frontally toward Spin – and could he have swerved, might it have seemed to Spin that he was changing direction? Finally, the five-step plan which de Waal hypothesizes Nikkie

LIVERPOOL
JOHN MOORES UNIVERSITY
AVRIL ROBARTS LRC
TEL. 0151 231 4022

might have had in mind (the suggestion is cautiously put) is keyed to a rapid-fire sequence of micro-events for which establishment of a correct order of succession is essential to an accurate interpretation of meaning. Thus, while again we rest our confidence in a primatologist's anecdotal relation of events put forward as evidence for intentional behaviour, again we remind ourselves of the close interconnection between interpretation of events and after-the-fact examination of their relationships. We would feel much more comfortable about it if we knew that de Waal had been able to check his recall of the precise order of events against a cine or videotape replay, just as we would feel more comfortable about Figan's quick glance from banana to Goliath (a highly persuasive detail in Goodall's account) if we knew that Goodall had been able to confirm her recollection of it by such means.

These are not niggling concerns. Anyone who has had occasion to test his or her recollection of events against film or tape records (as both Goodall and de Waal certainly have had occasion to do) will know just how unreliable memory can be. However, our reason for mentioning them is primarily to point out that they have to do with informational features which are available to participants in the actions outlined as well as to the observers. Human observers base their conclusions as to intention on perceived patterns of sensory input, that is, on those same relationships in time and space between grouped stimuli which in most cases may be assumed to provide the animals themselves with cues as to what is going on in their social world. Figan's monitoring glance at Goliath, which apparently alerted Goodall to the realization that there was something afoot, might well have alerted Goliath to a similar realization had he observed it. Nikkie, charging and hurling the brick, evidently did *not* betray to Spin his 'real' objective. Evidently, but it is hard to say, for events do not seem to have afforded Spin much freedom of response in any case – so that even when we accept de Waal's five-step explanation of events we are hard put to say whether what we discover is a strategy of deception or a strategy of containment. In our view, i.e., from an informational standpoint, it makes little difference; nor would it seem to have made much difference from Spin's point of view.

Recapitulating one of the criticisms to which anecdotal evidence is open is that one swallow does not make a summer. The main problem with de Waal's anecdote, as evidenciary material, is that the events related are complex, supercharged with information, and highly compressed, over in a flash (Nikkie's actions appear to have contained a great many cues, some contradicting others, a veritable swamp of social information, but a swamp moving at express-rate speed); and it is a sequence of events without compare in the literature of primate behaviour. Such a case can be documented satisfactorily only via cine or video camera, the virtue of which is that readers presented with a report that a swallow has been sighted can be confident (as can the reporter, his confidence bolstered by detailed analysis, frame by frame, of each micro-incident) at least that it was a bird. Documentation thus afforded, a set of 'moving image' data subject to analytic manipulation, is indispensable where such meaning-packed interactions are concerned. We do not mean to suggest that film or video can provide us with pictures of what is in an animal's mind; the point is to do all that can be done to document an action before taking that intentionalist short-cut into the mind's interior.

5. Sampling and related issues

A similar criticism of anecdotal observations of behaviour has to do with the fact that an aggregation of birds does not make a flight of swallows. This problem, of insuring that all cases in a sample are drawn from the same universe (a problem of formidable proportions where what is involved is behaviour of rare occurrence, variable in form, problematic in essential character, and variously described and interpreted by a number of different observers!) can be resolved to some degree, in analyses of intentional deceit, by attending to subcategories of behaviour within which comparison can be focused on formal likenesses which clearly signify underlying similarities of function and meaning. Construction of subcategories which are useful for comparison may be difficult or impossible: there cannot be many instances in which chimpanzees use weapons with intent to deceive, still fewer in which missiles are carefully aimed not to strike a target individual but to alter her or his direction of flight. However, examples are numerous in which observers report that primates, and chimpanzees in particular, have appeared to dissemble interest in food in the presence of others, only to snatch it up when left alone. Thus, Figan's case is by no means unique, though reasons for not manifesting interest (or for manifesting *dis*interest – which would seem to entail subtle but distinct differences in informational display) appear to vary in ways which have interesting implications for social cognition. If Goliath discovers the banana, tipped off by Figan's interest in it, he will take it for himself. In a similar situation involving captive chimpanzees, Menzel (1974) reported that a female 'leader' allowed to observe food being concealed in a one-acre field learned to put off recovering it, after release with companions into the area, until she could be sure that it would not be lost to a more dominant male – see also de Waal's accounts (1982) of similar behaviour by chimpanzees Dandy and Zwart at the

Arnhem Zoo. Where might makes right, a wise subordinate animal can sometimes increase his chances of obtaining a prized item by not attending to it too obviously; such 'not attending' may be both intentional and, functionally, deceptive.

Monitoring the regard of others

Where the prize at issue *is* food, and human beings are in control of allocation and use, an animal may be encouraged by human caretakers to consume only the food made available to it and not to interfere with consumption by others. Here might still makes right, but 'right' is defined by tradition peculiar to the captive setting and 'might' is a matter of sanctions imposed by human caretakers. In these circumstances, monitoring the regard of others becomes routine and probably is more frequently noted by human observers, not least because it is they themselves that are being monitored. Savage-Rumbaugh and McDonald (1988) argue that deception, rare in young chimpanzees, later on is practised with increasing frequency and sophistication. Around age 1,

> the first realization that others have to be watching you to know what it is that you are doing, appears. At this point, infants begin to attempt, when no one is watching, to do things that they have been told not to do. For example, Panzee [discouraged from drinking and eating from the glasses and plates of others] . . . began to monitor very closely whether this behaviour was being observed or not. If not, she would take someone else's glass from the table, drink out of it quickly and return it to the spot where she had found it, all the while watching the other party to make sure she was not seen. At 1 year of age, she could monitor only one person, no more. Thus, if a second person were in the room, she did not also check them to see if they were watching. By contrast, it did not matter to older juveniles if three or four people were in the room; they could and would monitor everyone present before doing something that they did not want to be seen doing. (Savage-Rumbaugh & McDonald 1988:228)

Witholding information and affecting disinterest

Doubtless chimpanzees, and other primates too, regulate their behaviour to withhold information concerning internal states (intentional or no) and at the same time monitor the regard of others to guard against failure. An early warning system may be defeated by adopting the same monitoring behaviour, as we ourselves may take care not to let on that we have seen anything amiss (see Savage-Rumbaugh & McDonald's account, 1988, of 'keep-away' games with Kanzi). At this level of stalemated action overt indications of intention may, understandably, be suppressed or 'leaked' in carefully managed scenarios.

The goal of competition and the coveted object of intentional deception need not be food; at Gombe, subordinate male chimpanzees become adept at a low-keyed style of courtship, wooing attractive females into the bush and away from the view of more dominant males (Tutin 1980). Figan in effect conceals himself from Goliath, in the incident cited above, in the same way and for equivalent if less romantic purposes. Withholding of information is basic to deception. It may involve affecting disinterest in a reward that is in view, not revealing interest (and, if necessary, concealing oneself as Figan is said to have done so as not to betray interest) in a reward that is not in view or has not been noticed, concealing oneself *and* the reward while enjoying it, or some combination thereof – Kummer (1982) has recorded a coupling of baboons in which both partners are concealed (behind a rock: she entirely, he sufficiently) from view by a high status male, suppressing give-away signals of pleasure enjoyed, a particularly nice example not just of brazen deception (assuming that Kummer's interpretation is correct) but of the effortful cost that concealing information may require.

6. Competition for material objects

Interestingly, competition for material objects, as opposed to competition for food, mating partners, and access to high status individuals, occurs but seldom, nor does it seem ever to entail high levels of aggression. Consequently, such competition may not give rise to deception. For young animals, novel objects may stimulate chase and contact games, in which possession appears to be defended primarily for the sake of the chase; but, for instance, stone handling macaques do not compete aggressively for the stones they handle (Huffman & Quiatt 1986; Quiatt 1986a), nor do termiting chimpanzees appear to compete for termiting implements. Chimpanzees which Boesch & Boesch (1986) have observed cracking open highly prized oil palm nuts did not appear to compete for hammerstones, which were in limited supply. In particular, status relationships do not appear to influence use. Boesch and Boesch's filmed documentation shows a high status male waiting patiently for a male of lower status to discard a hammerstone (the only one available in the clearing in which Boesch and Boesch had provided a supply of oil palm nuts for cracking) so that he can claim it for his own use (note our intentionalist phrasing). He gets his chance when the low status chimpanzee drops the hammerstone and walks off. Subsequently, that same individual returns and the scene is reenacted, this time in reverse, but in neither instance is there a hint of competition. Such is the case for males at least; it may be that female chimpanzees take a different view of hammerstone possession.

'Waiting patiently' puts the case too strongly. It is very difficult to resist the temptation to phrase what appears to be intentional behaviour in strong intentionalist language and thus move communication along. In this example, it is mainly the fact that when a hammerstone is discarded it is recovered immediately by a another individual which leads one to conclude that the reason that that individual has been standing by, relatively inactive, is because he was 'waiting' for the hammerstone. At least, and as far as we are able to judge from the filmed sequence, the subordinate male which *may* have in mind using the hammerstone when it becomes available appears neither to have his attention fixed firmly on it nor to be manifesting disinterest so obvious as to kindle suspicion regarding an underlying covetousness. Thus, it would be presumptuous to speculate, on the basis of demeanour, what he *might* have on his mind. But from the standpoint of social cognition there appears to be no reason for speculation, as long as the behaviour of the animal currently in possession of the hammerstone is not influenced by what is going on in the mind of the other. The point is that where we find it interesting to speculate about the intentional character of an animal's *social* behaviour it is (perhaps not obviously) because we deem an animal's intentions to have consequences for the fitness of others as well as itself – others who we therefore assume will be or should be interested in discovering what those intentions are. As Krebs and Dawkins have reminded us (1984), not only observers but animals observed are engaged in 'mind-reading'. Since what is important to those animals is not mental states *per se* but whatever behavioural cues may afford correct prediction of the outcome of interactions, we suggest that a phrase more appropriate (if less intriguing) and more meaningful from the standpoint of information analysis would be 'situation-reading.'

7. Tactical deception: learning to 'use' the social environment

While intentionalist terminology does not seem to us to constitute in itself an obstacle to 'objective' analysis of behaviour, it certainly feeds the does (s)he/doesn't (s)he controversy which, as we noted earlier, surrounds discussion of animal awareness, intelligence, and deception (see, e.g., the commentary on Whiten & Byrne 1988b). All deception, by definition, has to do with controlling the transmission of information. In the great majority of cases what is involved is inhibition (or concealment from inspection by companions) of response, from the momentary waver of gaze to prolonged whole body activity. De Waal (1982) reports a chimpanzee turning away under threat from another male until he can gain control over his fear grimace by forceful, manual reshaping of the mouth, another covering an erection with his hand in the presence of a high status competitor. Fewer but still numerous instances involve more active dissimilation in which one activity is 'pasted over' or substituted for another (Koko, reprimanded for chewing a crayon, signs *lip* and begins moving the crayon as though applying lipstick, Patterson & Linden 1981, cited by Miles 1986) or – much the same thing from a communications standpoint – a signal is emitted that is inappropriate to the context but useful to the signaller: non-human primates and animals of other species too have been observed to give alarm calls, apparently in the absence of any predator, and then seize for themselves food dropped or abandoned by a startled companion (see various contributions to Whiten & Byrne 1986 as well as accounts in Mitchell & Thompson 1986). In most if not all of these instances it is difficult if not impossible to say whether or to what extent intention may be at work; as yet there is no clear consensus among scientists on how to treat intention in the study of cognition and behaviour (see, again, the commentary appended to Whiten & Byrne 1988b, also the critique by Kummer *et al.* 1990 of intentionalist interpretations where naturalistic observations are concerned).

While there may be, and clearly are, problems with anecdotal evidence, the great body of anecdotal evidence accumulated in recent years for what has been termed 'tactical deception' (Whiten & Byrne 1986) leaves little doubt in our minds that chimpanzees, gorillas, and orangutans do employ tactical deception in manners similar to and toward ends not greatly different from those involved in human deceit. Similarities may be closer than is the case with other non-human primates. The problem as we see it concerns what it is that is gained for the analysis of behaviour by demonstrating (if it can be demonstrated beyond cavil) that animals other than ourselves act intentionally, also what it means to suggest that some animals act more intentionally than others. We have suggested *that the primary utility of those amassed anecdotes derives from their evidencing the degree to which dissemination of information is regulated in the course of social interaction.* We have already suggested too, in Chapter 4, that to facilitate comparison of cultural behaviour across species (i.e., for those who prefer an evolutionist to an essentialist view of human origins) culture may be defined as the means by which social processing of information is regulated. In the next chapter we will pursue this idea of culture as information-processing system and make specific suggestions as to how it can enrich anthropological understanding of human social life. But let us conclude this discussion of reflective awareness and purposeful action with a look at two examples, one anecdotal, one deriving from a controlled field experiment, which illustrate how experience can give rise to strategies for social action, and indeed for constructing the relational categories which make primate social life what it is. They are examples in which the observers themselves remark on the close relationship which exists between situation- (not 'mind'-) reading and the formulation of a strategy, in the first example, or, in the second, construction of a relational concept.

We return, for the first example, to the Gombe reserve and to Figan, in a continuation of that account by Goodall to which we referred earlier in this chapter:

> One day, when Figan was part of a large group and, in consequence, had not managed to get more than a couple of bananas for himself, he suddenly got up and walked away. The others trailed after him. [Goodall has just observed that resting chimpanzees are likely to follow one of their number who moves off smartly, without hesitation]. Ten minutes later he returned, quite by himself – and, of course, got his share of bananas. *We thought this was coincidence – indeed, it may have been on that first occasion. But after this the same thing happened over and over again – Figan led a group away and returned, later, for his bananas. Quite obviously he was doing it deliberately.* (Goodall 1971:104, emphasis added.)

The emphasized portion makes its point better than commentary can. That happenstance stimulated learning, and that learning involved reflection and the formulation of a deliberate plan, can hardly be proved. The literature of primate deception is replete with instances of apparent accidental learning, on the basis of one or a few trials, of behaviour necessary to the achievement of a goal, behaviour which is performed subsequently without hesitation to achieve similar goals (see, in this connection, Schiller's critique (1957) of Wolfgang Kohler's representation of chimpanzee 'insight' (1925). For instance:

> Chimpanzee Austin learns to stop his companion, Sherman, from bullying him indoors at night by going outside and making strange noises. He runs back in '. . . and looks outside as though there were something out there to fear,' whereupon Sherman grows fearful, 'hugs Austin, and stops the bullying'. (Savage-Rumbaugh & McDonald, 1988:28).
> Bonobo Kanzi learns that if, in a keep-away game, he appears to lose interest in the target object he may increase his chances of grabbing it,

which eventually leads to games with prolonged stalemates, the ball on the ground and, in Savage-Rumbaugh's description of the situation (Savage-Rumbaugh & McDonald 1988:233); both she and Kanzi feigning disinterest, 'both attempt[ing] by every move and glance to deceive the other in a game where deceit is the ground rule'.

Orangutan Chantek learns to exploit his real fear of cats to instigate searches for imaginary cats, when lessons become tedious. (Miles 1986:253).

Gorilla Koko discovers that when she blows bugs on her human companions she can be entertained by the 'shrieking and jumping' her behaviour elicits. 'Originally, she laughed at this outcome . . . now she chuckles in anticipation of the prank.' (Patterson 1986:943.)

And most or all of these subjects of 'ape language' studies learn to assign blame to others when the reward for an action turns out to be negative.

For proponents of creative thinking by the great apes, such an anecdotal gathering of observations collected singly over various times and places may at least be said to *resemble* a 'flight of swallows' – to return to that metaphor. As to Figan's intentions, Goodall adds a persuasive postscript:

> One morning, after such a manoeuvre, he returned with his characteristic jaunty walk, only to find that a high ranking male had, in the meantime, arrived in camp and was sitting eating bananas. Figan stared at him for a few moments and then flew into a tantrum, screaming and hitting at the ground. Finally he rushed off after the group he had led away earlier, his screaming receding in the distance. (Goodall 1974:103)

Our second example concerns those vervet monkey females observed by Dorothy Cheney and Robert Seyfarth, frequently cited as categorizing relationships and to which we have referred on more than one occasion in previous chapters. We describe here an experiment that involved playback of fear vocalizations by four offspring of different mothers (four trials each, for 16 trials altogether) on occasions carefully selected when the mother was located near two control females (thus 32 controls), themselves mothers.

> [While] . . . playbacks significantly increased the probability that both mothers and controls would look toward the speaker . . . [i]n general . . . mothers showed a shorter latency to look toward the speaker when their behaviour was tested against the controls . . . Mothers also tended to look toward the speaker for longer durations following scream playbacks than did controls. (Cheney & Seyfarth 1980)

Table 7.1.   *Vervet monkey recognition of mother–offspring relationships. (a) The number of control females who looked at another control, and (b) the number of control females who looked at the mother during the 10 s before and after playback. (c) The number of control females who looked at the mother excluding those experiments in which mothers approached the speaker*

Two-tailed McNemar test for small samples. (a) $P > 0.20$, (b) $P < 0.01$, (c) $P < 0.02$.

(a)

	After	
Before	−	+
−	19	5
+	1	7

(b)

	After	
Before	−	+
−	16	12
+	0	4

(c)

	After	
Before	−	+
−	15	7
+	0	4

+, looked at mother (or control); −, did not look at mother (or control).
After Cheney & Seyfarth (1980).

Results are shown in Table 7.1. What struck Cheney and Seyfarth as most interesting was that 'playbacks significantly increased the probability that controls would look at the mother, while there was no change in the probability that a control would look at another control.' Results are by no means unequivocal in their interpretation, as will be evident in examination of Figure 7.1. Three mothers moved toward the speaker on one of their four trials, behaviour likely to have attracted the attention of controls; when these most overt instances of contamination are separated out, differences (compare *a* and *c*, Table 7.1) are not so striking. The significance of these findings must also hinge on the assumption that at the beginning of each trial attentional structures along every dyadic pathway in these triadic groupings are alike, an assumption which seems dubious given what must have been an important methodological constraint in setting up playback situations, i.e., insuring that the mother of the infant whose scream was to be played back was in a suitable location relative to controls (with the relation between controls perhaps being secondary).

Cheney and Seyfarth are well aware of such problems as these:

> Although it seems unlikely that control females looked at mothers solely because of the mothers' relatively stronger response to playback, this possibility could not be ruled out entirely. Indeed, when evaluating the results presented here, it seems important to bear in mind the many factors that may affect the ontogeny of individual vocal recognition among primates. It is possible, for example, that adult females learn to associate particular screams with particular juveniles and these juveniles with their mothers both by hearing screams themselves and by observing the responses of others to those screams. (Cheney & Seyfarth 1980).

Not only is it possible, we find it difficult to see how learning of relationships could proceed except by ways such as this. This processing of information concerning the nature of mother-offspring relationships is in some ways much more interesting than Cheney and Seyfarth's conclusion (with which we agree), drawn from 'the fact that at least six control females looked at the mother after playback without any apparent cues from her . . . that vocal recognition on a truly individual basis is possible'.

What vervet monkeys do with their recognition of individual–individual relationships is perhaps the most interesting question. It cannot be assumed that Cheney and Seyfarth's control animals look toward mothers whose infants have screamed (in playback) out of simple curiosity, without expectation, i.e., reaffirming and reinforcing their understanding of categorical relationships. They might well be exhibiting interest in their own welfare along with, as we note elsewhere, some curiosity connected with expectations concerning role performance by the individual under observation. The implications of that possibility for research on non-human primate social cognition and social action are considerable.

8 *Knowledge in the social domain*

1. Introduction

It is easy, perhaps too easy now that we have grown accustomed to it, to speculate from a commonsensical standpoint about the competitive nature of animal behaviour and, speculating, to derive whole series of strategic and tactical principles from the single basic axiom of genetic selfishness. It is easy too, having speculated, to describe the overtly selfish, actively exploitative features of a social relationship, to note with regard to each interaction what one individual appears to be gaining from it, what another appears to be giving up, how the interaction may contribute toward increasing reproductive fitness of the first individual, how it may detract from that of the second. There is nothing wrong with this protagonist/antagonist approach; it has proved enormously productive for formulating hypotheses and seems both appropriate and effective as a way of *initiating* analysis of social behaviour. However, all analyses of social behaviour are partial, and if we remain satisfied with such an approach we are likely to end up de-emphasizing the role of reciprocity in relationships and devaluing the importance of behaviour 'designed' to maintain relational systems. For this reason, studies of conflict resolution (de Waal 1989b), resource sharing (Teleki 1973; Rodman 1988), cooperation (e.g., in predation: Goodall 1971; Boesch & Boesch 1990), reciprocal grooming (see Jolly 1985 for a brief review of grooming as tactile communication), greeting behaviour (Goodall 1968), and classificatory response to signals without apparent consequence in action (Seyfarth 1984) are important to correct a picture of primate behaviour which is still most often portrayed with emphasis on the agonistic accents.

The importance of such studies, however, goes well beyond correcting the picture. Positive, cooperative behaviour, conflict resolution, caretaking, and social maintenance activities are a good deal more than just the obverse and complement of agonistic behaviour; nor is it our concern in this chapter to 'explain' primate social behaviour in terms of something like dynamic balance between aggression and conflict resolution, some

184

point/counterpoint theory of the relation between agonistic and posi- tively affective or hedonic behaviour. We are concerned here, rather, with what knowledge may underlie primate social behaviour, what information bodies are employed and how information may be processed by individuals in society in ways which are not accessible to or not so easily managed by individuals leading lives which are essentially solitary.

2. Environmental and genetic information

Let us begin by recalling our observation earlier, in discussing animal communication and, more specifically, functions of signal reception which are not analysable strictly in terms of an actor's objectives in manipulating the behaviour of others, that from the standpoint of a signal receiver conspecific senders of signals can be regarded as detached transducers of environmental information. For the receiver such trans- ducers in effect extend individual sensory apprehension and expand the store of environmental information accessible to neural review (Quiatt 1984, 1988b; Small 1990; cf. Dawkins & Krebs 1978; Krebs & Dawkins 1984).

From this point of view, studies of conflict resolution, etc., which might be considered tangential to studies of primate communication *per se*, move us towards an understanding of why and how social relationships and information networks are regularly maintained over both the short and the long term. Maintenance activity, like communication, is of course a two-way process – it is for convenience of analysis, after all, that we adopt the perspective of actor versus reactor, receiver versus transmittor. Behaviour oriented towards social system maintenance reaches its fullest development in the language-dependent ceremonial greetings and rituals accompanying human exchange of food and material goods, information, and territorial or use rights (see Chapter 11). Attention to the system maintenance functions of behaviours such as these (and of homologous behaviours in non-human primates), as to their more narrowly conceived informational and communicative functions, can illuminate the processes by which social resources, like other resources, are integrated in use. This should provide a good start towards introducing into socioecological comparisons that dynamic analysis of behaviour which is essential to evolutionary explanation (Standen & Foley 1989a) and which Richard (1985) has suggested is largely missing from studies of primate behaviour.

The concept of information as signal *content* may not be absolutely essential to understanding how reciprocal maintenance of social bonds produces social relationships, but it clearly is useful, and an informational theory of communication clarifies nicely some relations implicit in the distinction drawn earlier between a (public) 'social domain of knowledge' and an (individual) 'domain of social knowledge'. We don't want that distinction engraved in stone; we can confuse ourselves if we think about it too long or too intently. It is primarily a way of sorting out on the one hand knowledge that an individual shares with other individuals, involving in some cases information which can be accessed properly only in the presence of others and with their assistance (see Chapter 4), and on the other hand knowledge which an individual may have which pertains to that sharing, to the social behavioural means of storing and accessing information, including information in the social domain of knowledge. We referred just above to a 'store of environmental information accessible to neural review', a rather awful locution but one which may nevertheless help this explanation along. By it we mean to distinguish information which is acquired primarily through experience and is transmitted via social interaction, within as well as across generations, from information which is transmitted biochemically across generations, by the genes. Perhaps it were better to say, given the problem of what constitutes a generation (i.e., for purposes of discussing communication between conspecific organisms 'within' and 'across' generations), that transmission of genetic information involves intermediate transfer by haploid agents, while information which is not gene-bound lends itself to direct transmission in interactions between diploid organisms (Box 8.1).

3. Repositories of information communicable via social interaction

Limiting discussion here to the latter, we can begin by making a rough distinction between three conceptually distinct repositories of socially communicable information, as in Figure 8.1. This categorization may seem most obviously appropriate to consideration of information which is processed by *human* culture and in *human* communication (including self-communion), but it is broadly applicable and will prove of particular use in making comparisons across primate species.

The first of the three repositories comprises *individual memory*. This is the store of information which any given individual may have in mind at some particular time, either in short-term (working) memory or long-term (deep) memory. Other individuals may have access to some or most, but never quite all, of the same or very similar information in parallel memory systems; so we can conceive of an inclusive larger store which is held in common by members of any group, a store to which each has access not according to need but to age, experience, sub-group affiliation, role, and status.

The second repository consists of this larger store and may be thought of as *working memory collective to a group* (cf. Carrithers 1990 on collective representations). What is in that collective memory and accessible to the group as a whole will depend in part on the group's size, age–sex composition, and social organization, for access to and inter-individual exchange of information stored in the culturally connected memories of individuals is constrained by all three of these group features.

The third repository consists of *extra-somatic stores of knowledge*. In the case of human beings it comprises notably those technology-dependent archives in which we store information assembled in aural, graphic, and (to lesser degree) olfactory images, i.e., books and libraries,

---

Box 8.1. Comparison of transmission features of genetic and environmental information

Within a species, information may be conveyed from one individual to another via genetic replication or sensory system activities. Genetic and sensory system 'communication' differ in a number of important ways, e.g.

Genetic information	*Sensory information*
Coded in biochemical units.	Coded in units of physical activity or pheromone emission.
Transmitted via haploid agents.	Transmitted primarily via activity of diploid organisms.
Transmission unaffected by direct feedback concerning the specific character of the message.	Transmission likely to be modified by feedback gained from observing the behaviour of 'target' individuals.
Transmission across generations only and restricted to chromosomal pathways between parents and offspring.	Transmission relatively unrestricted in terms of relational pathways; it may involve individuals of either or both sexes, most ages, and practically any relationship.

Figure 8.1.   Three repositories of environmental information: (a) individual memory, (b) collective working memory within a group, and (c) knowledge stored extra-somatically.

computer disks and tapes, punched cards (more frequently once than now), chemically coated film and electronically treated videotape, audio-tapes and records, sheet music, bottled perfume, dresser drawer sachets, etc., etc.

How shall we compare the cognition and behaviour of non-human animals with that of human beings in terms of these three repositories of information? There seems no difficulty in the case of the first two, though we need to remind ourselves that we know little about the precise functions of human language in accessing those repositories – however much language may be assumed to set our own species apart from others in this respect. It is the third repository which presents a problem, for the traditional anthropological preference for exclusionary definition of *culture* suggests a reluctance to recognize that other animals utilize this repository. We should not be in too great a hurry to rule out where non-human primates are concerned an extra-somatic repository of infor-mation equivalent to that in which we human beings file multitudinous artifacts of our language-dependent knowledge of culture. Chimpanzee sleeping nests, termiting sticks, debris from branch-waving displays, and stones and roots bearing the marks of palm oil nut processing fall into this category (see our discussion in the next paragraph but one). Clearly, however, parallels are closer in the case of the first two repositories, most obviously in that of the first, with the second more germane to present concerns and in some ways the more interesting.

We should note that of these three repositories the first is intra-somatic, the third extra-somatic, the second inter-somatic. As far as knowledge in the social domain is concerned, we probably should not hope to inventory it via some as yet undeveloped x-ray technique or, if it could be done, by spilling the contents out of container brains and counting them over. The character of knowledge in this domain must be determined in part by the particular relationships that obtain between individuals who 'share' it; bits of the information that make it up reside in those social relations, to be released in or expressed by specific inter-actions guided by social knowledge; cf. Carrithers (1990): 'Collective representations exist in and through . . . [social] relationships, and derive their significance from them.'

4. The extra-somatic repository of social information: a closer look

Let us take a closer look at the third, primarily extra-somatic, repository. Non-human primates do not have libraries, art museums, and computer memories. They have their own bodies, on which items may be draped in

play and scars inflicted evidencing agonistic engagement. More permanent artifactual evidence of cognition in action can be seen in displace-
ment of objects as well as in physical traces of use or modification toward
some instrumental application. A branch torn from a bush or a tree by a
displaying chimpanzee may be discarded on the slope below; stones
employed to crack open palm oil nuts may be left near an anvil root or
dropped out of a tree. Any attempt to summarize the informational
contents of this extra-somatic repository soon reveals how skimpy an
inventory is involved, not only in comparison with human beings. The
sleeping nests of chimpanzees and gorillas, for example, can hardly be
said to be laden with information when compared with the more permanent habitation of a robin or a cliff swallow, to say nothing of the
bowerbird's ornate, self-advertising construction; the hair-lined nest of a
ruffed lemur appears to contain as much *artifactual* evidence of cognitive
activity. This is not to suggest that extra-somatic information is of
negligible import in the lives of apes and perhaps monkeys. To a
chimpanzee the quantity and character of used twigs and grass stems that
litter a termite mound may be indicative of its productivity, and pulped
wads of leaf at the foot of a tree may suggest a water-filled hollow above.
For a young chimpanzee whose mother has just wiped its bottom with a
leaf, the stain of faeces on that leaf may argue its effectiveness as a wipe;
similarly the honey stick or termiting brush which he or she has watched
mother manufacture and use can be examined afterward as a kind of
refresher lesson; in these and no doubt in numerous instances relating to
harvesting and processing foods, discarded material may serve as little
handbooks of method.

By and large, however, extra-somatic storage of information is not a
striking characteristic of chimpanzee life in nature, still less of orangutan,
gorilla, gibbon, or monkey life (Figure 8.2). The technological features of
chimpanzee culture, if it is legitimate to make the comparison in those
terms (see Quiatt & Kelso 1985, from which Figure 8.2 is taken), still
involve primarily instrumental manipulation of material, with or without
the intervention of an object instrumentally employed, but with next to
nothing in the way of transformative manufacture. That observation in
mind, it is noteworthy that human culture, while it may involve a more
richly developed tool technology, is also language-dependent. Therefore
it is most useful to include modern *Homo sapiens* (including members of
contemporary hunter–gatherer and technologically simple agricultural
societies) in cross-specific comparisons where the object of comparing is
to show what a great difference language has made to human behaviour.

Where the object is to establish, insofar as may be possible, differences in behaviour which are *not* obviously dependent on transmission of information via language, it would be better to omit modern *Homo sapiens* from consideration and focus the comparison on earlier hominid forms, e.g., *A. afarensis, A. robustus, H. habilis,* and perhaps *H. erectus,* that is, on species in whom advanced language abilities remain so far undemonstrated and interestingly problematic. Conclusions concerning the language abilities of extinct species of course are profoundly limited by the few hints which anatomy and artifactual evidence afford (for partial reviews of the evidence from anatomy see Falk 1984; Wind 1989; Milo & Quiatt in press a,b; for an argument that artifacts of the lower palaeolithic yield little insight into language ability of their manufacturers, see Wynn 1989a). For these reasons, paleoanthropological reconstruction of (or speculation concerning) cognitive evolution remains for the present highly dependent on processual models derived from comparisons of living primates (Toobey & Devore 1987) and informed by research in computational cognition (e.g., Laird *et al.* 1986; for reviews see Pylyshyn 1984; Graubard 1988; Newell 1990).

To return to our third repository, which is exosomatic and primarily material, social knowledge may be stored in it as information embodied in artifactual remains of behaviour indicative of status, lineage affiliation,

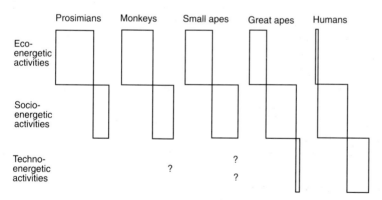

Figure 8.2. Energetic 'trends' in primate evolution: an approximate representation of energy expended by individuals in finding and consuming nutrients more or less in situ (eco-energetic activities), promoting and maintaining social bonds essential to reciprocal exchange (socio-energetic activities) and manipulating objects and tools and processing materials other than food (techno-energetic activities). Only the last two groups of activities, and especially the last, involve in significant degree extra-somatic storage of information. (From Quiatt & Kelso 1985.)

acquired expertise, and individual identity. This applies most obviously to artifacts of human behaviour, albeit in a highly limited range of stone and bone materials for those very early hominids which are most relevant to paleoanthropological comparison of the sort we have just referred to so pessimistically. Artifacts of non-human primate behaviour are, of course, similarly simple in material form, and, as the examples in Figure 8.3 illustrate, there are not a few of them: the nest built by a chimpanzee, the processed twig left by a chimpanzee that has learned to termite, the cluster of pebbles denoting stone handling by a young Japanese macaque at Arashiyama, and the dented kerosene tins used by a male chimpanzee (Mike) of the Gombe Stream reserve in noisy displays (Goodall 1971; Goodall has attributed Mike's concomitant rise in dominance status to these displays).

Information stored in the third repository is not embedded solely in durable material, as these examples may suggest. It seems appropriate that more immediate, highly transitory, artifacts of behaviour – visual, auditory, and other stimuli inherent in the act itself, 'images' as it were of behaviour in action – be categorized similarly as exosomatic and material. Seeing a termiting stick in use is at least as important as examining one which has been discarded, and the dents in Mike's petrol tins are not the same as the frightful din he has manufactured.

5. Storing and retrieving social knowledge
Let us see how human and non-human primates compare with regard to storing and accessing knowledge which is social not only in the character of its application but in the nature of its repository – and we remark first upon the well-nigh insuperable problem which faces anyone who would describe exhaustively the architecture and contents of any of these repositories. That difficulty recognized, and readmitting for purposes of illustration language-proficient human beings into comparison with monkeys and apes, let us consider now a hypothetical gathering of Appalachian musicians – or, more precisely, a gathering somewhere in the Ozark Mountains, at a time some years past, of local folk familiar from childhood with the music indigenous to their region, a limited and homogeneous suite of ridges, ravines, and hardscrabble hillside homesteads.

Someone, let us suppose, raises a song and is interrupted almost before he can get into it, after the first or second chord change, a line or two of verse: 'Well, now, that's not the way I learned it; the way I learned it, it starts out like this' – and demonstration follows, a rendition that may provoke another interruption and another demonstration, followed by still others if the song is one that is familiar in part or in whole, in one version or another, to many of those present. In this way the group may review quickly several versions of the same song, in entirety or in portions, arriving eventually at a composite rendition which all agree is satisfactory; or they may concentrate as a group on remembering just one variant which someone opines is original and 'right'. Or, conceivably, before any rendition is completed, the group may be distracted by one of its number who doesn't recall this particular song and who, furthermore, doesn't like what he is hearing of it. Saying, 'Now, I'm reminded of something just a little bit different, it starts off like this', he begins a tune that does not resemble in the least the one they

Figure 8.3   Four examples of social knowledge stored in the third repository (see text) as artifacts of behaviour: (a) chimpanzee day nest, (b) termiting tools, (c) cluster of stones handled by a Japanese macacque 'stone-player', (d) dented kerosene tins used in displays by chimpanzee Mike of the Gombe Stream Reserve.

have been fiddling with; and after a bit, seeing that he won't be stopped, the others join in, filing their versions of the prior song back in memory's store, along with or having incorporated variant phrasings which are newly mastered, i.e., with changes contingent on recognition of some error in recollection, on aesthetic preference for a particular variant just learned, on social influence, or on some combination of those and yet other factors.

This situation is illustrative of pedagogy in action, where pedagogy is defined as the transmission of an aesthetic standard (Premack 1984; see Carrithers 1990 for a discussion of the role of pedagogy in the evolution of human culture; both Premack and Carrithers consider the capacity for pedagogy unique to human beings). Although the example is hypothetical, constructed *ad hoc* for purposes of brevity and simplicity, the ethnographic record is replete with accounts of religious celebrations, life-passage initiations, ritual feasts and dances keyed to subsistence activities, in short, with descriptions of a variety of public ceremonies and informal gatherings which provide groups united in common goal-oriented activities with structured occasions for retrieving, assessing, updating, and restoring to individual memory knowledge in the social domain – in this sense of a repository of information which is shared by the group and which powerfully affects individual behaviour within the group, contributing more or less directly to coherent, persistent group organization. It may be arguable whether, for instance, a football spectacle or U.S. presidential debate should be considered such an occasion, or a stadium crowd or a scattered audience of television viewers such a group. We would argue that the informational content of a sports show is transmitted well beyond the arena of play, the social knowledge embodied in a presidential debate (to a far greater extent, perhaps, than knowledge that is in the narrow sense political) is recovered long after the television set is turned off.

6. Language and other forms of communication

Language plays a central role in these instances. As far as monkeys and apes are concerned one can only speculate, finally, about the nature of social repositories and the character of knowledge which resides in the social domain of knowledge. Some of the best examples of behaviour which appears to involve inter-individual transmission of information, e.g., chimpanzees communicating to one another the locations of concealed food (Menzel 1974) or baboons deciding, in assembly at the previous night's sleeping cliff, in which direction to set out foraging (Kummer 1971), are also amenable to explanation in terms of social facilitation, others in terms of manipulation and mind-reading. In instances in which chimpanzees have been observed to put off procuring food unnoticed by a dominant companion until that companion has departed (Goodall 1971; Menzel 1974), it is a nice question whether we should think of them as withholding information or as exercising care not to elicit through 'facilitation' or 'manipulation' behaviour which will work against their own interest. 'Mind-reading' (Krebs & Dawkins 1984) seems an awkward term at best for a theory of communication which disallows informational content; the scrupulously passive behaviour of animals actively engaged in *not* divulging what is on their minds argues content to be read.

'Mind reading' and manipulation

There must be times, and these may comprise most of the waking moments of a social animal's life, when an individual would *prefer* to have its mind read. Within chimpanzee communities, individuals in two sub-groups of temporary composition encountering one another after several days of foraging apart may display high excitement in interactions of dramatic intensity (Reynolds & Reynolds 1965). Moss (1988) has described similar greeting behaviour between individuals in groups of elephants coming together. The communicative function of such displays seems clearly to have to do with reconciliation and re-establishment of social bonds – the excitement may 'give ideas' to those participating of the intensity of pre-existing relationships (P. C. Lee, personal communication). But is the the energy expended on communication in these instances invested in information transfer or, alternatively, in trying to read minds and having one's own mind read? The line between social manipulation and information transmission at some point becomes so attenuated as not to be useful, if not here, then certainly at that moment in Sue Savage-Rumbaugh's study of the conjunction in a practical learning context of communication, tool use, and food sharing (see our discussion below), when Sherman, more assertive and apparently quicker to learn than his companion chimpanzee Austin, places his hand on Austin's and guides him to a proper solution to the problem which he has been presented (Savage-Rumbaugh 1986).

Sherman will receive some of the food with which Austin is to be rewarded for responding properly. Is Sherman then manipulating Austin's behaviour in a way that works to his own benefit? Yes, of course, and in the most straightforward way. Is he sharing knowledge with Austin? Yes again – and this instance belongs at least on the boundary of that collection of human essays at instructive manual shaping of behaviour, sometimes accompanied by verbal directions, sometimes not, that for at least some purposes it is useful to think of as informative (see, e.g., Gardner & Gardner 1969, 1973 and Fouts 1972). Sherman's 'shaping' of Austin's response appears to present much the same problem of ambiguity in interpretation as does shaping by human beings of chimpanzees' ASL or other gestural responses in 'ape language' experiments. Are those human beings to be regarded as trainers or as language instructors, and what's the difference? The main thing is that they are friends, partners at least, interacting regularly in a small society. The importance of that partnership aspect of ape language training, that is, of a close, continuing cooperative social relationship between trainer and subject can hardly be overemphasized. In the most productive of the chimpanzee experiments, i.e., the Gardners working with Washoe (1969, 1973), Premack with Sarah and other chimpanzees (1972, 1976, 1986), Rumbaugh with Lana (1977; Rumbaugh & Gill 1976), and Savage-Rumbaugh with Sherman and Austin and, subsequently, with the young bonobo Kanzi (1984a,b, 1986, 1990), language training sessions have been embedded in a rich context of social relationships, regular interaction with a few familiar trainers (in Lana's case via computer) and, probably significantly, routine transfers and exchanges of foodstuffs, toys, and other material objects.

Sherman's and Austin's world

A problem with the ape language studies has always been that the close mutual understanding which springs from social intimacy may be of such a personal and emotional character as to impede external assessment of the results produced (Terrace 1984). An exemplary case is that of Patterson, whose work with the gorilla Koko (e.g., Patterson 1978a,b, 1980a,b, 1981, 1986) deserves a better reading than many have been willing to accord it. The problem is inherent in the close relationship between teacher/trainer and pupil/subject, aggravated by the fact that that relationship is unavoidably central to the communications under study (Quiatt 1990). The extreme outcome of situations in which a communicative relationship grows in social importance while the world around it shrinks is privatization of meaning of the sort not infrequently observed between cranky masters or mistresses and their all too human pets or, indeed, between human beings themselves in relationships which are extremely close and extremely isolated – in which privatization of meaning may be reflected in a private language or in *folies a deux*. However, Sue Savage-Rumbaugh and her colleagues managed in a methodological *coup de main* to shrink the world around themselves and their chimpanzee subjects without allowing relationships within it to grow out of scale (Savage-Rumbaugh 1986). Their eventual goal was to teach Sherman and Austin to cooperate in obtaining food rewards and, in the process, to communicate symbolically with one another about problems encountered. But, since it was assumed that the basis for cooperation must lie in the expectation of reaping rewards to be enjoyed in common, and since the design of the experiment called for one individual to obtain and distribute the reward made available on success in a given trial, their trainers first had to teach these two young males to share food, which neither was inclined to do.

Neither chimpanzee showed much interest either in deferring gratification, which almost by definition is prerequisite to sharing. On the other hand, both chimpanzees *were* interested in play, enjoyed interacting with their human trainers, and liked simple games involving transport and transfer of objects. Savage-Rumbaugh reasoned that testing procedures rooted in an individual achievement/reward paradigm were not designed for training impatient young animals to share, but that Sherman and Austin might be trained into sharing with less frustration for all parties if, at mealtimes, individual servings were dispensed with and food distribution then and on every occasion of consumption be treated as a communal, festive event in which food exchange and consequent delay in consumption were handled as features of play. To accomplish this, all who worked with the two chimpanzees were encouraged to take meals with them and to share and exchange food themselves; thus, human trainers would be in a position always to promote the desired behaviour by example as well as exhortation. Similarly, Sherman and Austin were allowed to participate in preparation of meals, in unpacking and storing away newly delivered groceries, and in getting out and setting up food rewards (likewise the trays on which test items would be arranged, since testing sessions were by these means transformed into something like play-intervals within the larger food game). All this preparatory activity of course involved handling food which was eventually to be eaten. Sherman and Austin appeared to enjoy the activity partly for its own sake, not just as a way of getting at the food; so depriving them of the privilege of participating, when on occasion they forgot themselves, provided reinforcement of training – but it is important to see the preparation and the gamelike exchanges as the primary mechanism of training.

For an anthropologist, Savage-Rumbaugh's account of this transformation of a laboratory facility for the study of chimpanzee behaviour into a little world centred around the preparation and consumption of food for and by the chimpanzees and the people who shared it makes fascinating reading. The whole appears to have been carefully managed so that each activity, both in its character and in the means by which it is regulated, served a clear training function toward the specific goals of, first, persuading Sherman and Austin to put off eating food which was placed before them (without producing undue frustration – an important concern in view of the larger goal) so that they could be taught to share food, so that shared food rewards could be used to encourage cooperation and symbolic communication in the selection and employment of tools appropriate to the solution of problems submitted to them. As important as each individual activity, as each element in this complex but tightly crafted little behavioral system, seems to have been the system as a dynamic whole. In effect Savage-Rumbaugh constructed something like a self-contained interspecific cultural world in which all meaningful behaviour bore on the object of getting Sherman and Austin to employ a few concepts, some symbols representing them, and the objects to which they applied, in ways not so rigid as to suggest rote memorization of serial operations but conforming to patterns which would make sense to the human partners in the enterprise. Sherman and Austin (and Savage-Rumbaugh's assistants in training) were trained from the start away from behaviour that was not directed toward this goal. The meaning of the behaviour produced must in some sense reside in the goal toward which it is directed, but more interesting is that it appears to stem from social relationships (and component or contributive interactions), a material technology (including the testing apparatus), and a conceptual world, all of which centre entirely around food.

That is no doubt an oversimplification. Nevertheless, Savage-Rumbaugh's experiment evidences an important three-way relationship between language, social interaction, and materials processing. The last item in this tripartite relational system is commonly overlooked by primatologists concerned with animal communication in nature, perhaps because we tend to take the processing of natural materials by animals in their everyday life for granted except when we attend specifically to object manipulation and tool use in connection with feeding and other activities. The result has been a decoupling, in recent discussions of the evolution of primate intelligence following Humphrey (1976), of 'social' cognition from 'object' cognition. But culture feeds on nature, not just in post-modernist accounts of a capitalist Culture appropriating (and expropriating) Nature (see, e.g., Haraway 1989), but in a simpler, non-dialectical sense of structuring energy conversion; hence anthropologists concerned with an evolution of culture concomitant with that of cognition (if their concern is not solely with human Culture, the brand that is New and Guaranteed) are perhaps more likely than primatologists in general to be wary of such decoupling. In any event, in the experimental world 'reduced' by Savage-Rumbaugh to an elementary set of social relationships and material processing activities, a tripartite system emerged which, for her purposes, facilitated both interspecific and (for the chimpanzees) intraspecific communication via language; for our purposes it suggests that communication among non-human primates may be most richly informative (at least where human observers are concerned) where it affects relationships with material objects as well those with other individuals.

The blocks world

Compare Savage-Rumbaugh's experimental setting with that of SHRDLU, a computer software program designed by Terry Winograd to simulate the small world of a one-armed robot. In Winograd's description of this world, composed of variously coloured cubes and cones which can be combined into 'towers' and 'steeples' (see Figures 8.4 and 8.5), the robot

> . . . can manipulate toy blocks on a table containing simple objects like a box. In the course of a dialogue, it can be asked to manipulate the objects, doing such things as building stacks and putting things into the box. It can be questioned about the current configurations of blocks on the table, about the events that have gone on during the discussion, and to a limited extent about its reasoning. It can be told simple facts which are added to its store of knowledge for use in later reasoning. The conversation goes on within a dynamic framework – one in which the computer is an active participant, doing things to change his toy world, and discussing them. (Winograd 1973:154)

Here is a sample of dialogue, carried out in real time (with the responses of the 'robot' displayed on a television screen) between a person typing in commands and questions and the computer. The computer's replies are in upper case, and each interchange includes Winograd's description of what is going on.

> 1. *pick up a red block.*
> OK.
> The system answers 'OK' when it carries out a command. In order to pick up the red block, it had to clear it off by finding a space for the green one and moving the green one away. The initial situation is shown in Figure 4.1, and the result in Figure 4.2 [as shown here in Figures 8.4 and 8.5].

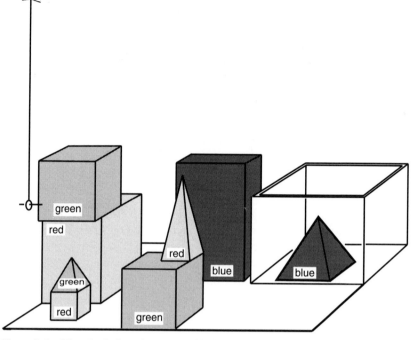

Figure 8.4. The physical environment of Winograd's Blocks World robot (Winograd 1973). See text for description.

2. grasp the pyramid.
I DON'T UNDERSTAND WHICH PYRAMID YOU MEAN.

There are three pyramids in the scene, and the program knows that phrases beginning with 'the' are intended to refer to a specific object the speaker has in mind.

3. find a block which is taller than the one you are holding and put it into the box.

The system has to figure out what is meant by 'one' and 'it', as well as using deduction to decide which block it is holding.
BY 'IT,' I ASSUME YOU MEAN THE BLOCK WHICH IS TALLER THAN THE ONE I AM HOLDING.

There were two possible referents for 'it': the one selected, and '*the one you are holding*'. The program has heuristics for choosing between them.
OK.

4. *what does the box contain?*
THE BLUE PYRAMID AND THE BLUE BLOCK.

5. *what is the pyramid supported by?*
    Earlier the system responded to '*the pyramid*' by saying 'I DON'T
UNDERSTAND WHAT YOU MEAN'. Now, it assumes that we mean the one
just mentioned
THE BOX. (Winograd 1973:155–7.)

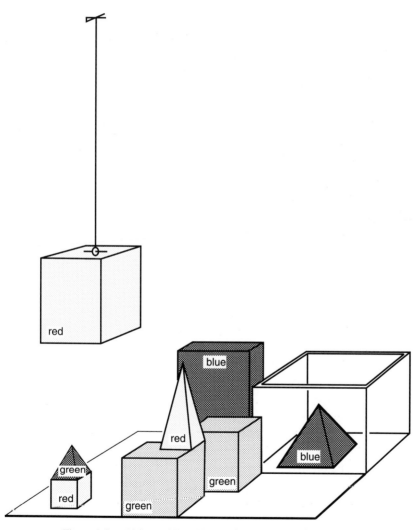

Figure 8.5.   'Pick up a big red block' (Winograd 1973).

The comparison is instructive in a number of ways. First, in terms of scale, small though the world of Sherman and Austin may be, the tabletop blocks world devised by Winograd is much smaller, not just physically but from the standpoint of activities afforded. We did not mean to imply, in describing the world of Sherman and Austin as experimentally 'reduced', that by developing a highly formal structure to shape connections between social interaction and food activity Savage-Rumbaugh has placed significant restrictions on the freedom of already captive animals. Sherman and Austin live in a comparatively rich captive environment, and there seems to us no reason to assume that behavioural satisfaction is in any way diminished through the imposition of the kind of structure we have described – even when accompanying food rewards are left out of account. Second, it would appear that Winograd's blocks world affords, despite its toy dimensions, a more richly communicative interactive exchange than is possible between Sherman and Austin and their caretakers – but this is a function, surely, of the fact that the dialogue between operator and computer is carried on in English and that the computer's partial but still relatively sophisticated understanding of semantically loaded english terms is conferred upon it by the program.

Margaret Boden, in a discussion of the contribution of common-sense knowledge to the interpretation of a linguistic expression on a particular occasion of its use, has compared the SHRDLU-programmed computer's response concerning what is meant by 'IT' to this exchange from *Alice in Wonderland*:

> '. . . Even Stigan, the patriotic Archbishop of Canterbury, found it advisable –'
> 'Found what?' asked the duck.
> 'Found *it*,' replied the mouse, rather crossly. 'Of course you know what 'it' means?'
> 'I know what 'it' means when *I* find a thing,' said the duck. 'It's generally a frog, or a worm. The question is, what did the Archbishop find?' [quoted from Boden, 1987.]

If the computer reveals more common sense than does the mouse (for it is the mouse who is obtuse here, not the duck, who knows what a duck's world is like but not as yet whether that knowledge can be applied to understanding the archbishop's world), still, it is not the kind of common sense that affords learning. From the standpoint of learning, though not self-evident in the brief extract cited here, Winograd's blocks world is deficient, not to say empty. Learning was provided for in a subsequent blocks-world program, HACKER, designed by Gerald Sussman (1975;

cited in Gardner 1985). Winograd's concern in constructing SHRDLU was to examine the dependence of language *understanding* (and accurate translation) on particular settings of use:

> 'We are always in a context, and in that context we make use of what has gone on to help interpret what is coming. Much of the structure of language comes from its being a process of communication between an intelligent speaker and hearer, occurring in a setting. The setting includes not only a physical situation and a topic of discourse, but also the knowledge each participant has about the world and the other's ideas.' (Winograd 1973:153)

7. Social relationships and communication

Winograd's and Savage-Rumbaugh's simplified settings for behaviour both afford insights into the relationships between semantic structure, social interchange between familiar individuals, and material/spatial features of context. Comparison is not intended to suggest (any more than was our discussion of simulation models in Chapter 4) that analysis of primate behaviour can be focused in future on computerized simulacrums. We do not mean to compare chimpanzees or for that matter human beings with machines (or iconic software);[1] the comparison is of contexts in which the character and quantity of information is deliberately limited so that acquisition, integration, and retention can be better controlled. Oversimplification to some degree is unavoidable in comparative discussion; the degree increases where comparison is with computational models in an inorganic world. A major difference between relationships in the artificial world of SHRDLU and in the strictly limited but nonetheless real world contrived by Savage-Rumbaugh is that there is no equivalent in the former to social relationships in the latter. This difference has important ramifications, for it is likely that in Savage-Rumbaugh's training world every 'relationship' which connects human beings or chimpanzees with food items and serving trays is coloured by those other, social relationships. However, differences between the Blocks world and Savage-Rumbaugh's small world of interspecific communication, and between both of those and that of the strictly human social group, should not lead us to discount similarities. In the SHRDLU world, in which activity boils down to translation of information received from outside, there *is* something like a social relationship between the operator and the program; its basis and for all practical purposes its sole attribute may be said to be communication about the nature of towers and steeples in terms of information available to both.

It probably is significant that communicative functions of a social relationship *across* species are rarely evident to external observers except in contexts of strictly limited and highly structured informational content. The 'conversations' which dogs initiate with their masters and mistresses most frequently have to do with acquiring food and going outdoors; many of the conversations which apes initiate with human trainers in experimental settings appear to be similarly oriented; the gorilla Koko, if we can accept at face value published translations of her reported sign-play, scathing denunciations of trainers, and insightful comments on life and death (Patterson 1980a,b, 1981), may be an exception. But, Koko apart, it would seem that where conversations between chimpanzees and human beings are concerned, neither partner appears to be particularly adept at assimilating information about concepts foreign to her or his way of life, however important those may be to an individual of the other species. In the absence of evidence there is of course little more to say about this possibility. We conclude, nevertheless, that a social relationship established across species, even between such phylogenetically close relations as chimpanzees and human beings, is unlikely to be as emotionally rich or to provide anything like an equivalent substantive basis in shared experience for mutual rapport and intimate communication of mood, motivation, and intended action as a social relationship between members of the same species.

Human beings do make fools of themselves in relationships with other human beings, often assuming shared conceptual understanding where none exists beyond getting the salt passed and finding a mutually tolerable thermostat setting. Not surprisingly then, human beings are as capable of deceiving themselves similarly with regard to relationships across species – where it is even less likely that harmonious interaction will be compromised by a dawning suspicion that conceptual accord is lacking. Still, no primatologist of our acquaintance is likely to suppose that his or her imagined or real rapport with individuals of another primate species comes anywhere near being as close emotionally and as quick and unambiguous in terms of informational exchange as can be achieved in relationships with other human beings. The key word here is *rapport*, for in either case there appears to be a natural link between emotional and informational features of communication; but within species the link is more intimate, an inevitable product of socialization within the family and local group of the developing individual.

Within species, social relationships of long standing provide the interactional vehicles for exchange of information (and material, notably food; see, e.g. Teleki 1973; Goodall 1986a,c), an exchange quickened by

Table 8.1. *Layering of roles and responsibilities in the mated pair relationships of human beings*

Role	Behaviour
Sexual	Mutual cooperation
Economic	Dual foraging strategies Differential technological expertise Reciprocal exchange of nutrients
Parental caretaker	Different contributions to diet of offspring and to subsistence training congruent with sex-based division of labour

From Quiatt & Kelso (1985).

emotional rapport. Where there is no language specifically 'designed' by natural and cultural selection for storing, accessing, and transmitting information, individual-individual exchange of information occurs in conjunction with regular close interaction between individuals long familiar with one another. In such relationships, exchange of information may be further facilitated by overlapping or layering of mutually compatible roles and responsibilities, as in Table 8.1.

8. Conventionalized knowledge and institutionalized social relationships

Here we approach the problem of how to describe 'institutionalization' of social relationships, a process which Hinde correctly sees as concurrent with the evolution of human culture (Hinde 1987). A reasonable inference is that language is somehow involved, and certainly, language or no, institutionalization of social relationships must involve conventionalization by some means of knowledge in the social domain. Perhaps it would be more accurate to suggest that the only way we can approach an understanding of the institutionalization of social relationships is through assessing and measuring the degree to which knowledge in the social domain has been conventionalized. Language does not appear to be necessary to conventionalization of such knowledge in its early stages, as shown by the responses of vervet monkeys to playback of vocalizations in those critical field experiments by Cheney and Seyfarth to which we have referred several times. We have described the experiments (Cheney & Seyfarth 1980) in which a distress cry emitted by a young monkey in an agonistic situation was played back later to a group of three females which included its mother, at a time when the youngster who had emitted

LIVERPOOL JOHN MOORES UNIVERSITY
LEARNING SERVICES

the cry was not in sight. While the mother responded as anticipated by looking toward the directional source of the sound, the first response of the other two females was to look toward the mother, presumably to see what her reaction was going to be. Here the possibility which interests us most is that monkeys in a group may have certain role expectations of one another, one being that a mother will assist her offspring when it is in distress. Without this inference the observation is in fact not very interesting, for it cannot surprise us that other animals perceive relationships between entities in their environment or that they should be attentive to disturbances which may entail risks to themselves. Specific role expectations apart, what is notable in this situation, in the response of the other females to the mother whose offspring's distress call is being played back, is the anticipation revealed by each female of a display by the mother of *some* index behaviour, some response in that complex of index responses that enables human observers too to recognize a mother–offspring relationship and give it a name, in the absence of genealogical information. The mother's reaction to her offspring's distress call, the monitoring responses of nearby females, and the human observer's assignment of a label, 'mother–offspring relationship', to what is being recognized – all are behavioural artifacts and informational indices to the conventionalization of certain bits of knowledge in this area of the social domain.

Conventionalization of knowledge in an adjacent area, as evidenced by the cooperative behaviour of Sherman and Austin, allows relationships to be extended on a basis which is not so obviously biological. Sherman and Austin displayed differences as individuals, Sherman being more assertive and quicker to experiment, Austin being diffident and perhaps more thoughtful – differences which Savage-Rumbaugh says remained stable through adolescence and which she indicates may be due in part to differences in early development; for instance, Sherman was removed from his mother (so that she could be bred again) at age 1½ years, whereas Austin was removed from his mother after only two months, 'having failed to thrive' in her care (Savage-Rumbaugh 1986). Whatever the reasons for Sherman's and Austin's differences in personality, those differences colour their relationship with one another and affect responses to training. When Sherman with controlled impatience 'shapes' Austin's behaviour by guiding his hand to a mutually rewarding solution of a problem which has been presented to Austin, we may think of that behaviour as having been shaped in its own way by a particular history of interactions on which their relationship is based, in short, by the character of the relationship and of the individuals involved. But we should not forget, either, that each of those interactions, considered from a communications standpoint, is conditioned by informational features of the moment – which perhaps is only a way of saying that Sherman and Austin's interactions, specifically here with respect to the solicitous laying on of hands by Sherman, cannot be considered apart from external features of the world they move in, which in this respect resembles closely the Blocks World. For some purposes – and consideration of the evolution of cognition in conjunction with conventionalization of knowledge in the social domain (and concurrent institutionalization of social relationships) is one of them – it is useful to conceive of both worlds as informational constructions.

A similar if less obvious example of cooperation in response to informational conditions imposed as a problem by a particular setting, in this case a more 'natural' problem even though the setting is one of provisioned feeding, and of the tendency for 'correct' behavioural response to influence in a conventional way social relationships, involves what appears to be division of responsibilities by habitual feeding partners in the rhesus monkey groups on Cayo Santiago, Puerto Rico (Quiatt *et al.* 1988, 1990). For three decades, the dietary staple for these monkeys, which range 'free' within a naturally bounded habitat to which their ancestors were removed some 50 years ago, has been commercial chow dispensed in metal feeders distributed over the island. The feeders are four feet long and double-sided, that is with four low bins extending out on either side, into which bins the chow rolls down from a high-rising central repository which is filled from the top. The bins, like that central repository, are lidded against the weather, with the lids being hinged at the back, so that a monkey wishing to feed must lift a lid to get at the chow or feed on chow that others have discarded. All monkeys show a marked preference for chow direct from the bins, but, while male monkeys evidently all learn to lift lids, some 40% of females go through life without lifting a feeder bin lid. It probably would be a mistake to explain this behavioural dimorphism in negative terms, i.e., some learning omission on the part of non-lifting females; it makes more sense to conclude that males learn a single feeding strategy while for females alternative strategies are open. It is by no means clear why a female should adopt the apparently self-defeating strategy of not lifting and follow it consistently throughout life, but if there is an explanatory pattern of behaviour it evidently begins with the close cooperation that obtains between lifters and non-lifters within *matrilineal families*, that is, within generationally-bounded groups of regularly interacting female relatives (the matriline operationally defined as a four-generation unit intermediate between a mother–offspring grouping and the extended genealogical matrilines documented by human observers) and their unrelated close friends and associates.[2]

In 1987, in the largest group of monkeys on Cayo Santiago (L-Group), there were 21 such units incorporating, potentially, sibships at the four adjacent levels of kin between great-grandmother and great-grandoffspring (with *ego*, the individual for whom kinship is reckoned in relation to whom kinship terms are applied, locatable within any of the four levels). This is neither as confusing nor as reduplicative in analysis as may appear, for none of the 21 units realized their potential in social interaction (not only because individuals die off as they age into the higher genealogical levels but because of increasing tendency with age, even in female-bonded species, for sisters to go their several interactive ways). Thus, ten of these 21 female-headed families contained only one adult female. Assuming what is in fact the case, that for females lifting is a strategy which is adopted, when it is adopted, at an age coincident with the onset of reproductive maturity, then by definition only a strategy of lifting *or* of non-lifting could be manifested in these one-adult-female families. As it turned out, five were lifter families, five non-lifter families. The remaining 12 contained two or more adult females (almost all contained either two or three) and so in theory each could pursue as a family group one of three feeding strategies: lifting, non-lifting, or mixed (with both lifting and non-lifting adult females. Two of them pursued a lifting strategy, the other ten a mixed strategy. These ten appeared to be operating in accordance with Richard Wrangham's suggestion (Wrangham 1980) that, in female-bonded (FB) species whose primary foods occur in a patchy distribution, closely related females are likely to cooperate in exploiting food patches. A tentative conclusion is that cooperation consists of a division of responsibilities, a lifting female providing access to food, her non-lifting companion(s) watching out for and initiating first aggressive responses towards intruding competitors. In these family groups, conventional delegation of role responsibilities was tied to biological relational status. However, there were some feeding partnerships in which a mixed strategy was exercised by mid-ranking males (who of course lift) and non-lifter females of low status. Similarly, the one old male who did not lift, presumably because his hands were badly crippled by arthritis, fulfilled non-lifter role responsibilities in partnership with the high-ranking female lifters with whom he consistently fed. Such data allow us to define fairly precisely what we mean by cooperation, i.e., enacting different but complementary strategies in synchrony (cf. Hemelrijk and Ek's distinction between 'reciprocity' and 'interchange', 1990); they suggest too that cooperative mixed strategies of behaviour displayed by unrelated as well as related partners in conventional (as opposed to biologically dictated) social relationships involve a division of role responsibilities with independent management of appropriate bodies of information. They suggest, finally, something which is most important from the standpoint of conventionalized behaviour and institutionalized relationships, which is that learning of behaviour patterns central to those relationships may for monkeys be extremely rigid, even in the case of behaviour (e.g., lifting and not lifting lids) in which we would have anticipated tactical flexibility and greater 'intentional' control over motor habits. Similar observations, broader and more general in scope, have been made in conjunction with analysis of some of the behaviour manifested by male and female 'friends' (defined operationally in terms of preferred grooming relationships and close spatial association) in free-ranging groups of Japanese macaques (Huffman 1987, 1991) and olive baboons (Smuts 1985; Strum 1987).

In human evolution, institutionalization of social relationships eventually involves language-dependent systems of reference and agreements as to what constitutes behaviour appropriate to age, sex, and various other status assignments. In monkeys and apes social relationships may be conventionalized in the sense that the behaviour of individuals involved in a relationship is not directly dictated by biology. Conventionalization of knowledge in such relationships involves individual management of joint and separate bodies of information relating to joint and separate strategies of action, the latter of which may be either reciprocal or non-reciprocal and complementary. Either way, conventionalization of the relationship appears to entail mutual agreement concerning goals which goes beyond rapport to involve transactional communication and probably, in addition, a pragmatic interchange of perspectives and 'self-objectification' of both parties along the lines suggested long ago by Mead (1934).

Conventionalization of knowledge in the social domain and of the particular ways by which information relevant to cooperative behaviour is accessed and managed is not dependent on language but provides the essential basis for further cultural evolution, including language-dependent institutionalization of social relationships. From the standpoint of the group as a whole, institutionalization of social relationships involves an increase both in number and in overall complexity of form, resulting in a greatly expanded variety of relational categories which are contingent on conventional recognition of kinship and other classificatory associations. Language, whatever its behavioural origins, brings this trend to full flower by affording conventional names for concepts and relationships as well as for material things (see Chapter 11 for fuller discussion of this point). For individuals striving to maintain and increase their reproductive fitness both in cooperation and, as occasion requires, in competition with their day-to-day interactional associates, precise delineation of social concepts, social relationships, social strategies, and social repositories of knowledge seems so obviously advantageous as to cause one to wonder why it is only in hominids that there has evolved a language which is extensive and flexible in its capacity for categorizing information.

A good answer to that question has to do with the exclusionary function of language; while it is clear that language affords extraordinary powers of cooperation for individuals who share its use, those powers can be directed with stunning effectiveness against species (as against conspecific individuals) who do not. For group-living hominids, language is a highly efficient tool for accessing, storing, and exchanging information essential to the construction and manipulation of marriage, mating, and social systems (Chapters 9 and 10). The capacity for language appears to be a species constant – although that is no doubt in some sense an article of faith, dependent at least on a definition of 'capacity' which is as vague as it is broad and general. The ability to *employ language as a tool* varies complexly among normal individuals in ways contingent both on biology and, perhaps no more today than in times past, on birth and opportunity. Language must therefore have altered considerably, for possessor (hominid) species, the context of natural selection – particularly, it seems reasonable to suppose, in the earlier stages of its evolution. The critical changes may have involved categorical expansion, contingent on the evolution of cognition, of information stored and, regardless of repository, readily accessible to working memory. When speaking of evolution consequent to those changes it would make sense to contrast not biological and cultural evolution but genetic and over-riding cultural selection as inseparable aspects of biological evolution.

Notes

1. Nevertheless, we do suggest that students of primate behaviour attend closely to developments in research on artificial intelligence, on the simulation of learning processes and on cognition generally. There are at least three reasons why primatologists should be interested in simulation studies. First, as we noted in Chapter 4, simulations of human cognition (at least those which are not focused specifically on language capabilities) generally are applicable to the behaviour of apes and sometimes monkeys as well as human beings.

Second, in certain areas of research, as in the study of behaviour in evolution, simulation may turn out to be the method of choice for many problems. Third, the study of cognition is an area of primatological research in which it seems likely that simulation techniques may on occasion be combined effectively with or even substituted for experiment. It is of course a worthy goal from a conservationist standpoint, where human ends and animal rights can be compatibly served, to substitute other approaches for experiments which put animal subjects at risk.

2. Quiatt elsewhere (1985, 1986b; Quiatt & Kelso 1985, 1987), in discussing problems of population structure and making comparisons with human social organization, has referred to such units as 'households', preferring *household* to the similarly problematic term *family* for defining, at this intermediate level of organization, sub-groupings which may include 'non-family' members. In contexts where conceptual discrimination is not a critical issue, substitution of terms may prove confusing, and so we dispense with it here.

9   *Kinship and marriage*

1. Introduction

In our search for the underlying continuities between non-human primate and human social action, we have come to focus on the transmission of socially relevant information between group members both during on-togeny and in the course of adult interactions. Unlike the society of other species, human (in particular our own) society is known to us emically, that is from the inside, and we consequently know that it is only as a result of learning that we come to differentiate and distinguish the social categories into which our fellows are located by us and locate themselves. The primary categories for every human being are those of his or her immediate family, and it is to these that we turn our attention in this chapter. The rationale for looking first at kinship (rather than friendship, say, or work-related roles) is that it is so elementary. Social anthropologists have almost without exception regarded kinship systems as the most basic of all social structures in human societies, and there is every reason to suppose that in evolution the formulation of kinship termino-logy and the conventionalization or formalization of relationships be-tween kin preceded other categorizations within society, though that assumption is incapable of proof.

At any given time, anthropologists approaching the question of the evolution or historical development of kinship systems have used particu-lar methods and favoured particular lines of argument. Fox (1979), for example, emphasizes the naturalness of kinship categories, and sees the elaboration of kinship and marriage rules as a wholly natural, evolved, biological process, the outcome of natural selection, achieved by the (equally biological) use of language. In his analysis, systems of marriage are outbreeding systems between sections or lineages of society. Hughes (1988) attempts to avoid the 'group selectionist' bias of such arguments, stressing the need to understand social behaviour in terms of 'compro-mise among antagonistic self-interests' (p.19). His approach, which is frankly sociobiological, is based on the principles of kin selection and reciprocal altruism, showing how the cooperation of kin groups and their relations with one another can be explained from these basic ideas.

Our own approach will be different. We have in previous chapters emphasized an informational approach to primate social behaviour, and in relation to kinship we want now to explore how already cohesive kin groups may have gained extra benefits by establishing links with one another. We do not see this simply as an extension of outbreeding, but as a means by which high status kin groups ensure a continuation of their status relationships with other kin groups, and as a mechanism for ensuring a regular supply of mates in the form of marriage partners.

We have seen the importance of intelligence in the analysis of non-human primate social life. Are the extra dimensions of human kinship systems simply a reflection of increased intelligence in humans? There are two reasons for caution. First, this might be putting the cart before the horse, i.e. the elaboration of kinship systems might have been the very selection pressure which brought about increases in human intelligence. And second, we find the greatest extension of kin lines, not in the apes, but in the macaques and baboons, so that degree of intelligence and ramification of kinship do not obviously coincide.

This seems to be an important point. Studies of how intelligence is used in social contexts have focused not merely on *ad hoc* manipulation of individuals but on the planning and execution of long-term strategies e.g. to improve social status or obtain a mating partner. De Waal (1982) has shown how intelligently chimpanzees manipulate both each other and objects to achieve social ends, and Dunbar (1984) has shown that the calculation by unmated male geladas of just when to launch an attack on a harem-holding male involves considerations of the size of his harem and how cohesive it is. Essock-Vitale & Seyfarth (1987) have reviewed studies on this issue, and Whiten & Byrne (1988b) have compiled and contributed to a review of evidence for tactical deception in primate social behaviour.

The above studies have a lot to tell us about how primate social behaviour compares with our own. They tell us we are not the only ones who make careful assessments before embarking on the establishment of social relationships, that we are not alone in being circumspect about whom we choose for friends, whom we can boss around and whom we had better be nice to. But the study of kinship is not primarily about the social psychology of relationships. Systems of kinship are not about dyadic relationships, they exist at the next integrative level, the level of social structure. Kinship does, indeed, have a psychological dimension. For example, the psychology of the mother–child relationship is very different in all societies from the psychology of the mother-in-law–son-in-law relationship. But the extension of kin relationships into matrilines (in primates) or matrilineages (in humans) takes us to the level of social structure.

In macaques, matrilines are ranked relative to each other, and the ranked matrilines of a group of macaques constitute its social structure. Likewise in the human case, the kinship systems studied by anthropologists are segments of the whole social fabric, often in traditional societies the most important segments, and the working of the structure of the whole society is only to be understood in terms of the interrelationships of the lineages of which the kinship system is made up.

2. Primate matrilines

Studies of ranked matrilines in monkeys have shown that the existence of these structures is important in determining the eventual social rank of individuals (Kawai 1958; Datta 1986), and furthermore that matrilines rise and fall relative to one another over a process of years. Samuels, Silk & Altmann (1987) describe two periods of rapid change of female ranks after 11 years of stability between matrilines in a group of baboons. In January 1983 there were violent fights, several females in one matriline combining against particular females in other matrilines, as a result of which 18 rank relationships between females were reversed. In the following August, three matrilines fell in status relative to the newly dominant one, involving downward shifts of status for 19 females. The fights were not as violent as in the previous January. The reasons for these reversals were not clear, but the authors suggest that matriline size, health of individuals, and the number of sexually receptive females were the most important factors involved.

Such inter-matriline fighting is unusual, and stability is the norm, but the above case does indicate well the underlying competition between matrilines. Normally, matrilines appear to be able to maintain their relative status, and the mechanism of how they do so is well known, having been first described by Kawai (1958). It was not until some time after the explanation of kin altruism by Hamilton (1964), when the theory of kin selection was applied to the explanation of primate matrilines, that their significance became clear. At the level of interactions, stability between matrilines is accomplished because high ranking mothers, by intervening on behalf of their offspring in disputes, are able to raise their own offspring's status relative to offspring of lower ranking mothers. A second process, suggested by Datta (1986), is that young members of a higher lineage seek out opportunities to dominate older members of a lower lineage, or, as opportunity presents itself, older members (including their own mother) within their own lineage.

Younger monkeys do this by waiting until one or more relatives are nearby, and then using these relatives as alliance partners to threaten an opponent. What we see in these cases is that a younger member of a dominant lineage enlists the support of kin against a non-kin rival. He or she is able to do this because of kin altruism. Kin are willing to support each other in agonistic situations, and this, we must assume, is a result of selection for behaviours which increase the inclusive fitness of one or other or both of those involved. In the present example, choosing the moment to act by the youngster, and support by the older kin, involve complex social learning factors. Fitness benefits are most obviously for the youngster, whose status is increased but it is also the case that all members of a matriline stand to lose fitness if the matriline itself loses status relative to another matriline. Thus fitness of matriline members is maintained by supporting kin, and can even be said to be increased because the status of a matriline is marginally increased every time one of its members raises its status relative to one member of another matriline. Such marginal-benefits may be very slight in themselves, but taken together over time they bring about the continued high status of a matriline and this results in better access to food and preferred habitat space (and hence reproductive success) for its individual members.

Macaque matriline structure

The structure of the matrilines recorded by Kawai (1958) is shown in Figure 9.1.

This figure is redrawn more or less exactly as Kawai first presented it. It shows a number of things. First, the age of the animals, second, parent–offspring relations, third, sibling relations, and finally whether alive or dead. (The latter are included for completeness of the record rather than because of any functional significance.)

Kawai was able to demonstrate in this paper what he called 'dependent rank', namely the rank obtained by an offspring which was dependent on its mother's rank and her presence. The technique he used was to throw a food item between pairs of young monkeys and observe which one took it; the result depended on the presence or absence of relatives nearby. There have been many further studies of the effect of kinship on the rank of offspring in Japanese macaques and other macaque species (e.g. Kurland 1977; Bernstein & Ehardt 1985; Datta 1986). All have tended to confirm Kawai's observations.

Our objective here is not to present further evidence for the existence of kin support in primates, but to probe deeper into the nature of

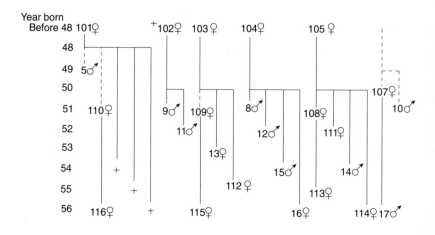

Figure 9.1. Macaque matriline structure. The members of Koshima troop and their lineage in 1957. (After Kawai 1958.)

relationships in monkey matrilines in order subsequently to make a proper comparison with human lineage systems.

The generalized structure of a monkey matriline, and the relationships in it, are shown in Figure 9.2.

Figure 9.2 differs in one important respect from Kawai's matriline diagram. In Figure 9.2 we designate one of the animals as 'Ego'. This is an anthropological convention, and it is particularly important when we come to designate members of a kinship system by particular terms. Such terms ('mother', 'sister' etc.) are applicable only in relation to a particular individual. It is also worth noting that analytic conventions of behavioural ecology involve calculating individual fitness from various

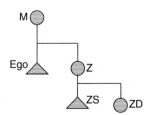

Figure 9.2. Three-generation monkey matriline. Key: M = mother, Z = sister, ZS = sister's son, ZD = sister's daughter. Horizontal lines above link relatives by siblingship. Triangles represent males, circles females. Dotted symbols show members of the matriline.

individual perspectives, i.e. the 'ego' perspective is not assumed, any individual can be 'ego' depending on whose interests are being analysed.

In the key to Figure 9.2 we state that horizontal lines above link sibling with sibling. This is the same convention as Kawai used. What is implied by these lines? In this case, the answer is based on biological relatedness. Ego's mother is his biological mother, his sister is such because she has the same biological mother as he has, his sister's son is the natural son of his sister and so on. The terms used are thus firmly rooted in biological relatedness, and if necessary the degree of relatedness can be calculated. For a sociobiological analysis of the pay-offs of altruistic behaviours, such mathematical considerations are important. For the present analysis no such calculations are relevant, but we need to note that the degrees of biological relatedness of members of a matriline normally correspond to their relative positions in the matriline.

The terms used do not *just* refer to genetic relatedness, however. Indeed, in the course of everyday life in the monkey group genetic considerations are irrelevant. What the human term 'mother' presumably *means* in the monkey world of Ego is the individual remembered as being warm, nurturant, supportive, and other such things. The monkey with whom Ego has a strong positive affectional bond is the one we designate on the diagram as 'mother'. Likewise, the meaning for Ego of the other relatives we have designated is based on personal experience of them during life. By the same token, non-relatives are monkeys with whom Ego has not had such interactions, and perhaps he may have had different, possibly stressful ones with them. As we have shown in earlier chapters, the background experience of relatives and others is stored as information in memory, and in the course of time, as a result of complex learning (the kind demonstrated in experiments on cognitive capacities – see Chapter 5), this experience is organized in categorical ways that enable monkeys to build up a comprehension of the relationships between other individuals in their group.

The kinship diagram is thus a representation of monkey *social structure*, displaying certain basic *relationships*, arising out of particular *interactions* (to use Hinde's 1976 terminology). And further, it is a representation of what, in a different, neural way, a monkey knows about these particular animals and their relationships with him or her, in its domain of social knowledge (Chapter 8).

The above points have been laboured somewhat because it is of great importance that we should understand what kinship in monkeys is, in order to avoid some confusions which arise when we come to make the comparison with human kinship. There are many implications of kinship terms when used of humans that do not apply to non-humans. For instance there are moral, legal and other cultural rights and obligations pertaining to the roles of kin in human families, some of which are certainly absent in the monkey case.

Let us take the term 'mother'. What is implied by the term as used in our own society, and how applicable is it to the monkey case? First, in our society, the term usually does designate a biological relationship. A mother is usually the person who physically gave birth to her sons or daughters. To that extent there is identity between the monkey and the human cases, and this is not a trivial identity, but is the foundation of the relationship in both cases. What else does 'mother' imply? It certainly implies a degree of emotional and physical closeness. It implies that the mother breast feeds her young offspring in the monkey case, and that she does likewise or bottle feeds in the human case. This is the first difference we have noted: monkeys do not bottle feed. Besides the feeding, a monkey mother holds her infant close to her, carries it around with her wherever she goes, and it clings to her by means of its hands and feet, under her belly as she walks along. In the human case we can see the bond is less continuous and, physically at least, less close. Human mothers in many societies put their babies down frequently into cots or cradles, and babies do not cling to their mothers' bodies. Mother–infant behaviour has evolved in a different way in humans.

Later these differences widen. The human mother surrounds the baby with her words and it grows up in a linguistic environment, that is in an environment in which information is encoded in words, not conveyed, as in other primates, by the organization and integration of observations and experiences into mental categories and distinctions. She clothes it and sees to its needs. She pushes it around in a pram and parades it in front of friends and relatives. She may be involved in an early religious ceremony in which the baby is incorporated into a religious community. A name must be given and recorded in a public record book. A future taxpayer has been born.

This is not at all frivolous. What we see here is the intrusion of three things into the human situation which are not so clearly present in the monkey one (but we shall check on this in a moment). First, the linguistic environment. Second, the intrusion of technology: nappies, cots, prams. Third, the incorporation into a socio-legal environment, by being recorded on forms and going through ceremonial induction into institutions. From an informational standpoint, all three involve increasing the variety and number of social stimuli by which we monitor and change environmental states. The human infant, child and adult are bombarded by vastly more items of cultural information to be learned and integrated into social skills.

Are there equivalents of these in non-humans? Certainly the maternal environment of monkeys is not a silent one. Cheney (1984) made an analysis of the ontogeny of vocalization in the vervet monkey, showing how early calls gradually change into the calls of adults, such as the set of alarm calls, which in adults are specific to the type of predator but in young animals are unspecific, and may even be mistaken. Thus Cheney writes:

> When infant vervet monkeys begin to give alarm calls, at about three or four months of age, they often make 'mistakes', and give alarm calls to species that pose no danger to them. What is interesting about these mistakes is that they are not entirely random. Infants give leopard alarm calls, for example, to many species other than leopards, but they give them primarily to terrestrial mammals (Seyfarth and Cheney 1980) . . . As vervets grow older, they increasingly restrict their alarm calls to the species that are most likely to prey upon them. (1984, p.6465).

The process by which this happens is not fully understood, but involves reinforcement from adults, and the mother is no doubt a significant reinforcer in this process. As pointed out earlier, we see here the development of categorical distinctions in primates, parallel to, if not homologous with, the categorical distinctions incorporated into words; in both cases the outcome is an information-rich store of relevant knowledge in the minds of individuals.

Regarding technology, this is a human *forte*, but its roots are certainly to be found in non-human primates. We have already referred to the incorporation by young chimpanzees at Gombe of the information required to make and use termite sticks, and in West Africa, of the tradition of breaking nuts with large stone hammers. Both these techno-logical traditions are transmitted mainly by females, and by mothers to their offspring. Clearly we must not exaggerate: the young primate is not *surrounded* by technology in the way the young human is. But a similar process goes on, albeit on a much smaller scale, and mothers are important in this process.

As regards the incorporation of the infant into the adult society, this most certainly goes on in non-humans, and observers have noted differ-ences between species and between groups within a species. As por-trayed in one of the vignettes at the start of chapter 2, langur monkey mothers pass their babies to other adult females for short periods within the first few days of life (Jay 1965). These females can thus get to know them and they are incorporated into the social life of the group. In macaques the same process occurs but later mothers are more possessive and there is a reluctance to allow other females to interact with infants until they are older. Evidence of the way primates understand the details of the incorporation of others in their group comes from vervet monkeys. The remarkable work of Seyfarth and Cheney has shown that at the age of two years, juvenile vervet monkeys are recognized by adult females of the group not just as individuals but as the offspring of particular females.

We have also seen that the presence of dominant kin close at hand has an effect on the interactions of others in macaque groups, showing again a keen sense of how individuals are incorporated and located in society.

Putting together the macaque evidence and the vervet evidence, it may be argued that monkeys, in these species at least, are very aware of something which we are calling 'kinship'. This awareness emerges in the context of rank acquisition, when older monkeys will give place to younger monkeys of senior rank if their mothers are nearby, and in the context of apparent danger to an infant, when its mother is looked to by other adult females for support. Both lines of evidence indicate that monkeys can have knowledge of what we would call the group's social structure, in particular its kinship structure. Indeed, in species where matrilines are important, it is *knowledge* of these matrilines, knowledge of who is related to whom and what that means in terms of social status, that makes the matrilines important. Without the knowledge, the matrilines would cease to exist. If monkeys kept forgetting who was related to whom, or never really learned this in the first place, there would be no matrilineal social structure, but instead the status relationships in the group would reflect the abilities of animals to dominate each other, either alone or on the basis of reciprocal altruism but not on the basis of kin support. Instead, we find ranked matrilines in which status is established within a year or two of birth in such a way that every member of a senior matriline has higher status than every member of a junior one, regardless of age.

3. Primate patrilines

In most primate species it is males who move from group to group in search of mates. As a rule, subadult males are more peripheral to group life than are females; eventually they leave their natal group to find mates elsewhere.

In a few species, however, it is females who emigrate. In chimpanzees living at the Gombe Stream Reserve, and at the nearby study site at Mahale, it is females who move out of their home community and seek a mate elsewhere. As a result, the genetic lineage, or core, of these chimpanzee communities consists of the males. Probably the same organization exists in the Kibale Forest (Ghiglieri 1984) and possibly also in the Budongo Forest, Uganda. We do not, however, find ranked patrilines, because there is only one patriline in each chimpanzee community, judging from existing studies (Goodall 1986a; Nishida 1979). The males in a chimpanzee community are very tight-knit. They tend to move around together, to share mates with each other, and it is they who form the hunting groups which kill prey species such as monkeys, bush pigs or small antelopes, and share the meat after a kill.

Motherhood and fatherhood
We have examined motherhood in some detail already. Fatherhood is a more complex topic. Nearly all primate species studied show evidence of male–infant relationships of a friendly kind, such as play. In some species, such as marmosets, males carry their young habitually, but this is uncommon. In most species, males carry infants occasionally. In baboons, males establish regular grooming relationships and association-al ties with particular females and these females subsequently become preferred mating partners; in due course when infants are born the males defend them and take on protective roles towards them (Smuts 1985). Figure 9.3 shows the structure of an olive baboon matriline, in which the adult females have semipermanent consorts, who are also shown. Males take less interest in infants whom they could not have sired, i.e. who are born shortly after a relationship has been established with a female. This should not, however, be taken to indicate that they understand the relationship between sex, pregnancy and paternity; more likely there are proximate factors at work, and it has been suggested that care for an offspring by a male is very often consequent on a long-term relationship with its mother (Smuts 1985). Stein (1984), in a very detailed study, draws particular attention to another aspect of male–infant interactions. Following a lead provided by Deag & Crook (1971), he documents the

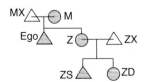

Figure 9.3. Olive baboon matriline. Key as for Figure 9.2 with additionally MX = mother's mate, ZX = sister's mate. Lines joining symbols centre to centre represent mating relationships.

LIVERPOOL
JOHN MOORES UNIVERSITY
AVRIL ROBARTS LRC
TEL. 0151 231 4022

occurrence of 'agonistic buffering' in baboons, by which is meant the holding of infants by adult males at times of tension and potential fighting. Such behaviour very often protects the baby-holder from attack and is therefore a potent reason why adult males should 'befriend' females and their offspring, whether their own or not.

In some species paternity has become an issue of great importance. This is seen most clearly in the case of infanticide, for example in langurs of the species *Presbytis entellus*. Sugiyama (1965) first described an attack by an adult male on a group of langurs in which he systematically killed all the infants. This phenomenon has been recorded many times by different observers (e.g. Mohnot 1971; Hrdy 1977a, Vogel & Loch 1984; reviewed by Hausfater & Hrdy 1984; Struhsaker & Leland 1987). In areas where infanticide occurs, langurs live in one male groups consisting of a male, his 'harem' of females, and their young. He keeps out all other adult males. Such males form 'bachelor bands' who live without access to females. However, from time to time such unmated males launch an attack on the one-male group, and attempt to drive out the male. If he can do so, one of them drives out the other and becomes the breeding male, and it is he who then kills the females' small infants. The result is that they resume oestrus fairly rapidly, and he is able to replace the offspring of another male with his own. As long as he can bring them to maturity (by resisting other takeover attempts) he has succeeded in transmitting his genes to the next generation.

This may not sound very much like the paternal care we mentioned in the case of baboons, but it is no different in principle. Just as the male baboon takes more care of an infant if he has had a long relationship with its mother, and it is therefore likely to be carrying his genes, so the langur male is achieving the same objective in his brutal way. Both cases demonstrate that fatherhood, like motherhood, is primarily based on the fact of genetic relationship between parent and child, though what proximate factors are organizing the behaviour is much less clear. One thing is certain, however, namely that in both cases, however different, the male concerned has time to familiarize himself with the social situation he is entering into; he collects information and builds up an organized store of knowledge, which provides the essential basis for his subsequent actions.

In a few species of primates, such as gibbons, fatherhood is relatively uncomplicated (though even here this is not always the case – see below). For simplicity's sake let us take a straightforward situation. Gibbons are monogamous (perhaps serially monogamous), and thus the social group theoretically, and often in practice, consists of an adult male, an adult female and their young. Both parents defend the young, and take equal shares in doing so, primarily by emitting loud vocalizations at dawn and throughout the day, establishing their ownership of their family territory (Carpenter 1940: Ellefson 1968). In addition the mated pair engages in vocal duetting, which according to Brockelman (1984) serves to strengthen the bond between them. Paternity confidence is high in such a case, and the male does all which is necessary for the support of his female and their offspring. At puberty, however, the young male gradually withdraws to the periphery and subsequently leaves the group.

Complications may arise because of demographic changes caused by arrivals, departures or deaths of males. Thus Brockelman & Treesucon (1986; see also Quiatt 1987a) noted that of three young males in a family group, the older two were in fact brothers of the current resident adult male, while the youngest was the female's offspring by that adult male's predecessor. Gibbon groups may not always be as regularly constituted as is sometimes assumed.

Dispersed lineages and the functions of brothers

Clearly when group size is as small as it is in gibbons, with a mated pair and their offspring very often forming the social group, there is no scope for the development of matrilines or patrilines as functioning, intra-group entities. However, there is always the possibility that such lineages do exist as *dispersed* entities. If this were the case, then an unmated male might, for example, be able to join up with a mated brother on the latter's territory. At present, despite a few reports of adult males attaching themselves to a mated pair (Chivers & Raemaekers 1980), we have insufficient information on the genealogical background of most hylobatid groups.

The question of dispersed lineages is a potentially very important one. It has been argued that, on Cayo Santiago, a subadult rhesus male leaving his natal group is likely to emigrate in the company of a brother or brothers, or end up in a group in which a previously emigrated brother has established residence (Drickamer & Vessey 1973; Meikle & Vessey 1981; Boelkins & Wilson 1972). The argument for brotherly influence on either timing or direction of migration at Cayo Santiago remains unsettled; Colvin (1986) maintains that 'there is no evidence that . . . [young males] emigrate in the company of their brothers'. However, there is good evidence from species other than rhesus that males transfer with peers (potential siblings) or into groups with brothers/natal group members already present (P.C. Lee, pers. commun.). If brothers do prove to exercise any influence then we can see another very important function of matrilines: they serve to further the mating interests of males within the lineage (see Quiatt, 1987b). Whether this is also true for females in the case of chimpanzees is not known, but it would seem probable that a migrant female chimpanzee would find it easier to enter and settle in a community which had already accepted an older sister than in one where she had no prior contacts.

We have by now seen something of the nature of mother, father, brother and sisterhood in primates. We have shown how some kind of understanding exists between kin, that is, animals who would normally be genetically closely related to each other and who have been in close association with each other during the growing period. Such animals support each other in a number of ways against non-kin, and their special relationships are often continued late into life, or even until death, as in the case of chimpanzees where an ailing mother may be accompanied closely and groomed by an offspring.

At the psychological level, there is much in common between kinship in primates and kinship in man. Continuity here is not hard to see. However, we have already in the preceding discussion spoken of social 'structure'. What is meant by that term is the way in which the (psychological) relationships between animals standing in the relation to each other of kin become, when considered together, the principle of organization of the society as a whole. The coherence of a rhesus monkey group on Cayo Santiago results from the sets of relationships within and between its matrilines. If and when the group splits, this usually happens between matrilines, one or more hiving off from the rest. This does not mean that matrilines are static, we have already stated that their strength waxes and wanes, and their rank relative to each other changes over time.

4. Human comparisons

We can now ask: do the lineal structures of non-human primate groups bear comparison with their equivalents in human societies? We think that to some extent they do, more especially in smaller scale societies. For instance, in the Kachin of highland Burma described by Leach (1977), lineages distinguish themselves clearly from each other, and a most important part of the distinction is relative status, with chiefly, aristocratic and commoner class lineages adopting particular behaviour patterns towards each other.

Leach writes: 'Matrilateral cross-cousin marriage plays an integral part not only in maintaining this class structure but in defining the 'feudal' relationships between chiefs, headmen, and commoners. The two most general principles that govern the Kachin marriage system are that a man will do everything possible to avoid marrying into a class beneath him, and that a man will seek to make the maximum profit either in terms of bride-price or political advantage out of the marriage of his daughters.' (1977:84).

The Kachin marriage system is complex, but the principles of maintenance of rank and of lineage recruitment emerge clearly. A feature of the Kachin system is the way lineages are incorporated into the political structure of the society. This is achieved by marrying a daughter or daughters 'down' a class, leading to control at the subordinate level. Leach illustrates this by a diagram portraying an idealized situation comprising 'three chiefs one of whom has under him three headmen, one of whom has under him three commoner local descent groups' (p.85). We reproduce this diagram below in Figure 9.4.

Not only in Kachin lineages, but in lineages everywhere, individuals are encouraged to maintain the lineage's rank, and members do what they can to keep up the status of fellow members. In patrilineal societies fathers can be most helpful and influential in maintaining the status of their offspring, in a great variety of ways. Sociologists distinguish between 'achieved' and 'acquired' status. The latter is what we are now talking about, the status that derives from the position one is born into in society. Brothers and sisters, too, can help one another to succeed by sharing information, social contacts and resources. For instance, paralleling the case of the migrant rhesus monkey, migrants coming into a more affluent country from a poorer one in the hope of improving their financial position can be greatly helped to get started by the presence of one or more kin who have already settled in the new environment. As in primates, human lineages can operate when they are dispersed. Indeed, they can do better, for they can operate in this way *even if the kin have never previously met one another*. This is achieved through language use and in particular *naming*, to which we shall return.

Marriage and fatherhood

We have now reached the point at which we need to make a preliminary analysis of a human kinship system in order to make a more systematic comparison with the kinds of systems we have looked at in primates. We took as our paradigm case for primates the monkey matriline (Figure 9.2). If we were to consider this as a human family, we should rapidly notice a very odd thing about it, namely the fact that both the mother and her daughter are single mothers, that is they are unmarried. This is quite

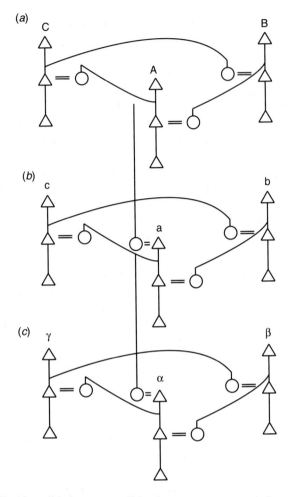

Figure 9.4. Kachin political structure. Triangles represent men; circles, women. (a) Chiefly class: three chiefs in mayu/dama circle form a loose political federation. (b) Aristocratic class: three village headmen together with chief, as fourth village headman, control political affairs of domain. (c) Commoner class: senior members of commoner local descent groups, together with village headman, as representing his local descent group, control affairs of the village. (From Leach 1977.)

a usual feature of families in western society these days, but in world-historical terms it is most unusual. More conventionally, in all societies the world over, and especially in societies in the less affluent parts of the world, childbearing is preceded or accompanied by the action of marriage, a characteristically human institution, which legitimizes a partnership between a man and a woman in such a way that any children born to that woman, whoever the genetic father, obtain through that marriage a legitimate place in society as the heirs and descendants of one or the other or both of their legal parents. This may at first strike the reader as a curious definition of marriage, but it is nevertheless an accurate one, and follows the definitions of other anthropologists, e.g. Mair (1972:91). Marriage takes many different forms in different societies. The rights and obligations it confers on the husband and the wife differ widely from society to society. Marriage may be considered a long-term or a short-term arrangement. It may be exclusive, as in monogamy, or not, as in polygyny and polyandry. Where there is some emphasis on patriliny, the wife may change her name to the husband's, and her children may take that name too, so that she and her children are lost to the lineage of her birth. In matrilineal systems, the children are likely to take the clan-name of their mother and their mother's brother.

Just as Figure 9.3 represented a three-generation matriline, Figure 9.5 represents a three-generation human lineage. To make the comparison as clear as possible, we have chosen to show a matrilineage.

Here, as before, the categories of individuals (M,F,ZH, etc.) are as seen from Ego's point of view. There are several departures from Figure 9.3. Note that parents are now designated in a different way. In Figure 9.3, the male partner of a mother was designated as her mate, and this was shown by joining the symbols for the female and male concerned by a line which ran centre to centre. In Figure 9.5, by contrast, the male partner of the individual labelled 'mother' is called 'father', and the male partner of the individual labelled 'sister' is called the sister's 'husband'. These terms are precisely used here to designate affinal kin, kin through marriage, and as such mark a departure from anything found in the animal kingdom. Because these are marriage relationships and not mating relationships, the individuals concerned have been linked in the diagram by horizontal lines below, a standard convention in anthropology. Lest there be some confusion over the word 'father' (which is an ambiguous term in ordinary usage), it is here used specifically to mean

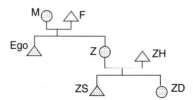

Figure 9.5. Three-generation human matrilinage. Key as Figure 9.2 with additionally F = father, ZH = sister's husband. Horizontal lines below link spouses.

pater (the legal and jural status of father) not *genitor* (the biological father). This is not to deny that the *pater* may in many cases be the *genitor*, but to emphasize the analytic distinction between the two.

This distinction gets to the root of the difference between non-human and human kinship systems, and it raises a problem. If we are going to quibble about the use of the term 'father' for human beings on the grounds that it is not the same as 'mother's mate', then what about such terms as 'mother', 'sister' (and by implication 'grandmother', 'cousin' etc.)? So far we have used the terms 'mother' and 'sister' of humans and nonhumans alike, but it could be argued that this muddies the very distinction we are trying to draw, between jural and biological relationships. This is not, however, the case. The important point we are trying to make is the extent of continuity between non-human and human kinship systems. It is only in some relationships, notably those created by the institution of marriage, that we need to distinguish between jural and biological relationships. In other cases, we can use the same terms for humans and non-humans because jural aspects are simply added on to the basic biological relationships. This applies, for instance, to the terms 'mother', 'sister', 'grandmother' and 'cousin'.

Fortes (1983), like Wilson (1983), has pointed to the incorporation of the father into the family through the institution of marriage as the single most important innovation marking off human social evolution from anything found in the rest of the animal world. 'Fatherhood', he wrote, 'is a creation of [human] society . . . man's central rule-generating institution' (1983:20). Fatherhood is here identified as the father figure, the mother's husband in patrilineal societies and the mother's brother in matrilineal ones. In Fortes' analysis, like that of Fox (1980) and Ingold (1990), human society is sustained by systems of rules that have juridical and moral force, are backed by authority, and are accepted and changed by consensus. The father, in Fortes' scheme, is the central disciplinary element in the process of rule transmission and enforcement. Fortes did not consider the argument for the equivalence of human and primate kinship to be valid, but he did not provide an in depth analysis of the primate data of the kind provided here. His analysis of the emergence of human society nevertheless is very profound and worthy of serious attention; but it is characterized by a wider gulf between the human and the non-human than we ourselves accept.

We have noted above that it is often the case in primates that the male who is the current mate of the mother takes a protective interest in her offspring, and it is this interest which is safeguarded and formalized in humans by the marriage ceremony, in which he takes on specific obligations and accepts that his heirs will have specific rights. But before discussing the differences between mating relationships and marriage relationships, let us remind ourselves of one respect in which marriage and the incorporation of the father it achieves seem to have similar characteristics to the process of mate selection in animals. In many, perhaps most, human societies, marriage partners are not allowed to come from the spouse's natal lineage. There is here a close parallel with the situation in most animal societies, where *mating* occurs outside the natal group, one sex migrating for the purpose of seeking a mate. The explanation usually advanced for the evolution of out-mating in animals without regard to proximate considerations is the selective advantage resulting from the avoidance of inbreeding depression. Marriage is not the same thing as mating, a point to which we shall return, but marriage very often does involve mating and may be the institution in which most mating occurs. Exogamy in humans thus shares certain features with out-mating in non-humans.

From mating to marriage

We have noted that there is an important difference between mating (a sexual relationship) and marriage (a jural arrangement which usually contains a sexual relationship). This seems to mark a major divide between animals and humans, and we need to consider why this might be so. What are the factors which might have brought about the institution of marriage in humans? The term 'marriage' includes a large variety of customs and is based on different ideas in different cultures. We do not intend to review these, but rather to take some of the most striking features of marriage and try to draw out its distinctive features when compared with primate mating patterns.

A useful starting point is the process of mate selection. Marriage involves among other things a mate selection process based on social knowledge. In some societies, marriage is organized by the marrying pair, but in many this is not the case, and selection of a suitable spouse is in the first place done by the parents. Depending on the extent· of parental control, economic and reproductive factors will be foremost in the selection of a marriage partner. Where such control is great, future spouses have little or no physical contact before marriage and sometimes find they are incompatible afterwards. In our own society, sexual compatibility is regarded as an important prerequisite for marriage. In arranged marriages, the most important consideration for the parents is the social status of the family of a prospective spouse. In much of rural India Hindu fathers with offspring who might make suitable spouses for each other meet and discuss with each other the possibilities of a match, and a major consideration on both sides is the amount of dowry payment by the father of the bride to the family of the groom. Among African cattle-herders, payments often go the other way, in the form of bridewealth paid by the father of the groom to the family of the bride. One of the factors determining which way the payments go is the relative evaluation of the 'value' of a bride as against a groom, and this is determined by many factors, such as how much work the bride is expected to do for the family of the groom, and whether marriage is uxorilocal or virilocal (Murdock 1949; Mair 1972; Radcliffe Brown & Forde 1950).

For present purposes we need not consider the matter of which way payments go, nor the amounts and kinds of payments, nor the details of evaluation. We need only note that marriage is a corporate process, operating between families or lineages rather than between individuals. This is one of the features which marks off marriage from mating. *Individual* men and women find mating partners in the community in which they live both before marriage and after marriage, and any given man or woman may have dozens or hundreds of mating partners in a lifetime. By contrast, the same individual may not marry at all, or may do so once or a few times. Mating is ordinarily a part of marriage, though not necessarily so, as in the case of the Nayars of S.India in which very young girls were ceremonially married to warriors but never had sex with them (for a few other marginal cases, see Murdock 1949), or in the small number of cases in our own society and others in which a marriage takes place but is never consummated. These are interesting because there are rules in some societies about the legitimate dissolution of a marriage if one partner fails to consummate it by the act of sexual intercourse (Reynolds & Tanner 1983).

Marriage is therefore something very different from mating, and not to be confused with it. Why then marriage? We have already argued that kinship systems are, in part, systems of knowledge. Affinal kin, those acquired through marriage, add to the network of people with whom one is involved in terms of specific rights and obligations, and in so doing they widen the field of relationships, adding to social knowledge. So much is this the case, that in traditional village communities individuals may be related by descent or affinity to almost everyone else in the community, and in some more inbred cases they may be related to each other in several ways.

This situation is in marked contrast with that which prevails in a monkey group, where matrilines, even though they can act together (e.g. in group defence), are more generally in a state of competition with each other, and are ranked relative to each other. Recall the process by which an individual relies on matrikin to determine his or her status. The outcome of intra-lineage solidarity is to homogenize the status of lineage members, marking them off clearly as a group from the members of other lineages. In this situation, competition between lineages persists. By contrast, marriage reduces the competition between lineages, and has often been (and still is) used to cement alliances between them.

We should not forget that competition between human lineages can be fierce. The institution of the feud is an example. In common parlance one talks of family feuds; many English villages, for example, have families who will not talk to each other, each family is solidary, each revels in malevolent stories about the other, marriage between the two is frowned on and consequently rare. But feuds can be much more devastating than this. In the early history of Iceland as related in historical accounts such as Njal's Saga (Magnussen & Palsson 1960), insults or bad behaviour by a member of one family to a member of another could, if unchecked, lead to violent confrontations and revenge killings, reprisals being taken by kin of the losing party over lengthy periods of time. In this acephalous social system, kin reciprocity and support was the basis of social order.

Reverting to the situation in primates, kin solidarity is well attested. That formation of aggressive coalitions is determined in part by kinship has been especially well documented in macaque species, though we have evidence from langurs as well (see Figure 9.6). Matriline conflicts in macaques have, for example, been described for *M. mulatta* (Kaplan 1977, 1978; Chapais 1983), *M. fuscata* (Kurland 1977), and *M. radiata* (Silk 1982). The evidence is overwhelming for macaques, and conclusions probably can be generalized to include most species in which females remain in their natal groups and take sides in quarrels. As Silk (1987) states, females 'form coalitions on behalf of maternal kin more often than on behalf of other individuals'.

There thus seems to be good evidence of the fact of lineage competition in both the monkey and the human cases. A number of social anthropologists have drawn attention to the fact that marriage is a potent method for bringing about alliances between potentially inimical lineages, a point that is nicely expressed by Mair (1972:87) when she writes that the people studied by anthropologists often say 'we marry those with whom we fight.'

Figure 9.6.   Langur matrilines in conflict. S. Blaffer Hrdy/Anthro-photo (see Trivers 1985).

5. Group selection

If we conclude that humans developed the institution of marriage in order to reduce inter-lineage conflict, then we are moving in the direction of a group-selectionist model of human social evolution and this requires justification. The problem is this: given that competition normally takes place between individuals, and that kin solidarity does not challenge this because kin are genetically closer than non-kin, how can we explain inter-lineage solidarity when by definition different lineages are genetically different?

Allen (pers. commun.) has pointed out that for social anthropologists there is some ambiguity here in the word 'kin'. Bilateral kin are closer to ego genetically than non-kin. But lineages are uni-lineal. Mother's sister's daughter belongs to the matrilineage, mother's brother's daughter does not, though they are genetically equidistant. Thus members of different lineages may be genetically equidistant kin. Nevertheless, lineages defined collectively do have greater within-group genetic kinship than exists between lineages.

The group selection answer is put in terms of group solidarity. We envisage competing groups, each group consisting of a number of lineages. Any group which has developed mechanisms between its lineages so that they are able to support each other will have advantages in fighting against groups in which lineages are opposed to each other. The latter will tend to fragment under pressure from larger, more solidary groups, will be driven from the best habitats, and their members will have reduced fitness as a consequence.

This is not unreasonable. It has been argued (e.g. Alexander 1987) that the explanation for the evolution of human institutions of many kinds is the pressure on whole groups of the practice of raiding, fighting and warfare, which led to the evolution of mechanisms of group cohesion, group survival, and brought about a reduction of individual and lineage competitiveness in the human case in favour of sacrifice and altruism for the group.

In actual fact, the extent of group action as against individual or lineage action is very much a response to socioecological constraints. At times when lineage competition is not of any consequence, individuals can be expected to behave in a self-interested way. When lineages come into competition with each other we can expect to find lineage solidarity, and when groups come into conflict we can expect lineage differences to be swamped in favour of corporate group action. These processes have all been documented in non-human primates. The latter, for instance, was documented by Southwick, Beg & Siddiqi (1965) at a temple site in North India. At this site there were a number of rhesus monkey groups.

> If a subordinate group failed to see a dominant group approaching, with the result that two groups came into sudden and unexpected contact, a severe intergroup fight would occur. 24 severe intergroup fights and numerous minor ones were observed in 85 days of observation. All these resulted from two groups coming into sudden contact owing to the failure of one group to retreat. Normally the adult males began the fight, but females and juveniles also became involved. These fights were ferocious and dangerous to the monkeys, often resulting in severe wounds, and most adult males bore wound scars around the face, shoulders or rump. Wounded individuals were fewer among rhesus in rural habitats and forest areas, suggesting that the crowded conditions of temple and urban environments resulted in more aggressive activity. In forest areas more space and protective cover greatly reduced the number of intergroup contacts (pp.142–3).

The existence of group solidarity, such a prominent feature of human tribal groups, is thus on a continuum with the group solidarity seen in primates, even though primate groups are not endogamous in the way human tribes are. The rhesus example given above is not atypical but commonplace (e.g. Holloway 1974). Fighting between groups has been reported in a great many species. Normally the context is competition for a valued territory or other scarce resource (e.g. Dittus 1986), but not always. The group conflict which occurred between the Kasakela and Kahama chimpanzees at Gombe Stream Reserve, in which the males of the Kasakela group exterminated, over a period of months, the males and even a female of the Kahama group, was not clearly about the control of food resources but may have been about access to females, since the Kahama group was a splinter group which had broken away from the Kasakela group not long before the fighting started (Goodall 1986a).

We have emphasized macaque or langur inter-group conflict as an example of groups acting together *as groups*, ignoring intra-lineage loyalties in favour of joint corporate action. Thus any argument that marriage arose as a way of uniting lineages in order to strengthen group action must take account of the fact that primates can act as a group without marriage links between kin lines. Group conflict is thus unlikely to provide more than a part of the explanation for the emergence of institutions linking up human lineages. Let us look at another kind of argument for the emergence of marriage, this time in terms of recruitment.

6. Recruitment

Levi-Strauss (1949) pointed out that lineages very often associate with particular other lineages, establishing mutual exchange of marriage partners, and he showed how the common institution of cross-cousin marriage could achieve this (see Figure 9.7)

The exchange portrayed here was, Levi-Strauss assumed, following Durkheim's much earlier insights about social solidarity, a primeval breakthrough in the organizing of society, associated with the elimination of incest by the introduction of systematic exogamy. The emphasis in that explanation is not so much on group selection in a context of inter-

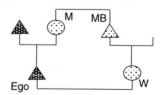

Figure 9.7.   Matrilateral cross-cousin marriage. Key: Ego = the marrying male, W = the woman he marries, M = mother, MB = mother's brother. Lineages are patrilineal.

group hostilities as on the organization of intra-group mating partners into a sustainable pattern of long-term reciprocal relationships.

Something similar to this has in fact been reported in non-human primates. There is evidence of selective mating between particular matrilines and particular patrilines. This comes from a study by D. G. Smith & Small (1987) who studied mating preferences in three captive groups of rhesus monkeys at the California Primate Research Center. These groups were separately housed in field cages (0.2 ha), so that the normal pattern of male emigration seen in wild rhesus monkeys could not occur. As a result, each group contained both patrilines and matrilines. Smith and Small were able to show that females from particular matrilines chose males from particular patrilines as mates more frequently than chance would predict, and this was true in each of the three groups studied. The results were not an artifact of the age or sex constitution of the groups concerned, but were the outcome of selective favouring of males from particular patrilines by females from particular matrilines.

In the wild patrilines are not co-residential owing to the pattern of male emigration at puberty. Females can thus only favour males from particular patrilines if they are able to detect patrilineality and if appropriate males are available, i.e. if they come into the female's group. Nevertheless, it has been suggested that matrilines on Cayo Santiago, where male emigration occurs, demonstrate a higher degree of genetic homogeneity than would be expected from random mating with incoming males (McMillan & Duggleby 1981). Thus it may be that even in free-ranging conditions females are selective about the lineage of the males they mate with. Perhaps, indeed, they prefer to mate with males from particular other *matrilines*. Either way (i.e. matriline or patriline preference) the result would be an increase in genetic homogeneity of the female's patriline over levels arising from random mating or mating with any male who happened to enter the group.

These non-human primate data are of interest for a number of reasons. The benefits to matrilines from this kind of selective mating as seen by D. G. Smith & Small (1987) are in terms of lineage solidarity arising out of increased kin selection, which is in turn a result of the increased genetic homogeneity of the lineage. An increase of genetic homogeneity of the same kind would inevitably result from the human institution of spouse exchange between particular lineages, whether matrilineal or patrilineal. The question arises, however, whether this resulting increase in genetic homogeneity would be reflected in an increase in cooperative, kin-selected behaviour in the human case (such a question could also be asked of the primate case). In social insects there is no doubt that genetic homogeneity leads to cooperation and we can speak of kin selection. In primates the situation is more complex because kinship behaviour is a product of closeness during development rather than shared genes. Nevertheless, whatever its basis, kinship and lineality do seem to be rather fundamental features of social life in insects, primates and humans, and it does seem that the preferences for mating between particular primate lineages constitutes a parallel with the human situation in which certain lineages exchange spouses. Whatever the origins or causes of these situations, and even if they are rather different, there may be functions in terms of lineage solidarity to such arrangements, and a further function which we shall now explore, namely continuity of lineage recruitment resulting from privileged access to particular mating or marriage partners.

The way in which a cross-cousin marriage system (ideally) works over the generations is shown in Figure 9.8. The important thing to notice about Figure 9.8 is the lines connecting D and WBS and connecting S and WBD. These lines indicate the reciprocal nature of cross-cousin marriage in terms of the exchange of wives between lineages. In one and the same generation, the son of the left-hand lineage is marrying the daughter of the right-hand lineage, while the former's sister is marrying the latter's brother. (It is of interest that Charles Darwin himself was a product of cross-cousin marriage.) This exchange is sustainable down the generations. It can take the form of symmetrical bilateral cross-cousin marriage. Or it can be asymmetrical patrilateral cross-cousin marriage as illustrated here. Or it can be asymmetrical matrilateral cross-cousin marriage. Whatever the particular circumstances, and these depend on the system of descent of the society as a whole, each lineage involved in cross-cousin marriage stands to gain from the arrangement, quite apart

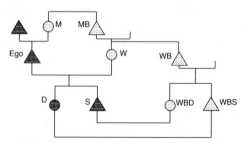

Figure 9.8.   Wife exchange between two lineages. Key: as in earlier figures with additionally: D = Ego's daughter, S = Ego's son, WB = Ego's wife's brother, WBD = Ego's wife's brother's daughter, WBS = Ego's wife's brother's son.

from (and possibly in opposition to) any gains to group solidarity. *There is no reason why the group should gain as a whole.* These two lineages may be a pair of aristocratic lineages which are keeping more than their fair share of the group's resources to themselves. They might in fact be impoverishing the whole of the rest of the group. We cannot know in any particular case unless we investigate the facts. The process of wife exchange between lineages can, however, provide a structural basis for social solidarity at the group level.

The exchange of women back and forth between a pair of lineages tends to be ideal rather than actual. Demographic factors at any given time may make it difficult to find an exactly appropriate partner for any given man, and so a well-known phenomenon comes into play, the fudging of kinship to allow the rules to be followed (Hiatt 1968; Chagnon 1989). Levi-Strauss was aware of this when he wrote that 'rules have a life of their own', in other words they are at times more honoured in the breach than in the observance. This is a very interesting point for our analysis. When relationships are fudged in this way, say when a Yanomamo man is looking for a spouse and has no cross cousin but the rules demand one, then he will consider girls of appropriate or younger age who fall close to the desired relationship, and by devious twisting and reinterpretations of relationships between himself and the girl will conclude that she is indeed a cross-cousin (Chagnon 1989).

In his account of kinship knowledge among the Yanomamo, Chagnon (1989) showed how detailed this was. Local groups were saturated with kinship links and ties, as well as opposition and rivalries. Chagnon was interested in sociobiological interpretations, focusing on the way that 'big men' could accumulate wives and thus increase their reproductive success by manipulating kinship links. From our perspective this is a case of manipulating information, restructuring social knowledge. Once a kinship system has come into being, individuals who can learn its rules and master its complexities are in a position to use it for their own ends. Our analysis of kinship systems as systems of knowledge lends itself to the idea that clever individuals (or parents, in the case of arranged marriages) will follow marriage rules which are of benefit to them, but bend the rules if conformity threatens to leave them without a marriage partner.

If we are right in interpreting the evolution of rules of cross-cousin marriage as a solution to the problems of lineage recruitment and lineage survival, then such rule-bending makes very good sense. Although the chosen spouse is not, technically or biologically, a cross-cousin, she is nominally one, and lineage satisfaction is assured on both sides; the game can go on. We are here looking at the history of social structure, we are seeing how the rules go on and the structure goes on, changing slowly over the generations as a result of progressive deviations from previous norms. Within this process individual human beings achieve more or less reproductive success, but bending the rules is clearly a different process from that of cultural change. Bending the rules may or may not produce cultural change; cultural changes involve transmission and survival of the rules themselves. In providing each other with spouses, lineages achieve a reliable source of recruitment of future personnel, fudging is allowed for emergencies, and the lineages go on.

From the time that an arrangement such as cross-cousin marriage is instituted between two lineages, it is in the interests of each lineage to look after the welfare of the other, because each is responsible for providing the other with the wherewithal to continue its line. Cross-cousin marriage seen in this way is a product of biological evolution. It is the result of a cognitive structuring of society which has brought with it selective advantages to the members of the lineages engaging in it. Murdock (1949), in his survey of the world's cultures, found that 'under unilinear descent . . . cross-cousins . . . are the commonest objects of preferred primary marriages' (p.320). The arrangements ensure reproductive opportunity for most lineage members, even if they do not guarantee reproductive success. There will still be competition between individuals, especially in societies where polygyny is permitted, so that despite lineage solidarity and marriage links between lineages some individuals will still fail to obtain a spouse, and some will be infertile. To this extent, the rules and structures of kinship systems and systems of affinity, although manmade, themselves provide the setting in which competition, now conventionalized, takes place and in which differential reproductive success is achieved. It is in this sense that we can talk of humans living in a world of their own construction, but also in one in which (at least until the advent of modern affluent contracepting societies) the normal processes of natural selection take place.

Figure 9.8 presented a two-lineage linkup. In societies with moiety organization this may be all there is. Individuals belong to one of two clans, and the clans are exogamous. This kind of organization is not common however. More often there are many lineages or clans in a society. Depending on how many there are, and precisely what kind of symmetrical or asymmetrical marriage links there are between them, the society may be closely knit as a unity, or there may be divisions such as those of class or caste. Models of interlinking marriage systems can be proposed and compared with actual situations. In a circular model, if lineage A exchanges spouses with lineage B, B with C, C with D and so on, and in the end the last lineage, N exchanges back with A, then a perfect form of social cohesion of the entire group will result. In real life, such systems are inevitably imperfect in operation.

Names

We have written so far about lineages and the importance of lineage recruitment without saying much about what human lineages actually do or what they are for. A lineage is a *descent group*, it has to do with the matter of descent down the generations. Let us look at the situation in our own society. Here, each descent group has a family *name*. This name is passed from father to children, i.e. patrilineally. Property, however, is inherited from both parents equally. Thus rules of inheritance are bilineal. Some property, notably rings and other jewellery, are often transmitted from grandmother to mother to daughter, i.e matrilineally. We can thus see from our own society that transmission of intangibles and of tangible objects is a complex matter, and social anthropologists have shown that the ways in which the rules governing how all the items, statuses, titles, honorific roles, rights, duties and physical objects are transmitted from generation to generation vary from society to society.

In our society and in many others, the name of the family is in some respects its 'identity' and as such is the nearest thing to its real 'essence'. To besmirch a family name may be a punishable offence, or one which can lead to feuding. The reason is surely that the name is a very powerful evocation, encompassing all the family members, past and present, at home and far away. In some respects, the name *is* the family. This intangible marker is the lineage indicator, and its mode of transmission marks off patrilineal from matrilineal descent groups. The extent to which it, or its equivalent, matters to the lineage members and to others is a measure of its importance, and of the importance of matters of lineage and descent in the society concerned. In Western society, its significance is somewhat less than in most societies in the less developed parts of the world, where much of the economic life of the people is organized along kinship lines. In some societies, a person is to a great extent defined by the kinship position he or she occupies, though this is a very complex issue (Carrithers 1985). In Iceland, a person has two names, a first or individual name and a second name which states of whom he (she) is the son (daughter). Thus Magnus Magnusson, the well-known Icelandic author and TV personality, is Magnus, son of Magnus; had he a sister called Gudrun she would be called Gudrun Magnusdottir. The Rekyavik telephone book makes interesting reading!

7. Conclusions

Let us draw this chapter to a close by reviewing some of the conclusions we have reached so far about the similarities and differences between primate and human kinship systems. We noted first of all that the study of kinship systems, whether in humans or non-humans, is a study at the level of social structure, not at the level of dyadic relationships or interactions. This is an important point, because it introduces into the discussion supra-individual processes, going beyond single-lineage considerations to considerations of the whole group. Our understanding of the nature of human social structure itself is that it consists of institutions which are in turn the products of human inventiveness, and are transmitted down the generations as systems of knowledge, the knowledge of individuals being subsets in the social domain of information available generally in the society.

We looked at the mechanics of ranking of individual members of monkey matrilines, and saw how the outcome consisted of a series of ranked matrilines in species such as macaques. Intra-matriline cooperation and even altruism can be understood on the basis of the close genetic relationships between relatives in the lineage. With this we contrasted the mechanics of human lineages, in which the actions of individuals towards each other are to some extent based on the rights and obligations vested in their roles. Nevertheless, we saw that there was a considerable degree of overlap at the biological level between primate and human lineages and the relationships within them; in the human case the biological relationships are overlaid with sets of jural rules. We also saw that a *preoccupation* with kinship was characteristic of both primates and humans.

We then moved on to a consideration of the exclusively human institution of marriage. This, and the consequent incorporation of the father (*pater*) into the family, have sometimes been seen as the most significant steps in human social evolution. We examined the similarities and differences between mating arrangements in primates and human marriage. Despite the fact that the latter creates a jural, not a biological relationship, we saw that there are overlaps between the two in terms of mate selection and care of offspring. One main function which we found marriage was performing was in the linking up of lineages, and we concluded that its primary function at the level of social structure was the organization of an efficient system of recruitment to lineages. We noted in particular the corporate nature of the marriage process, and saw how in smaller societies the whole group could be unified by marriage links between lineages, thus reducing the dangers of group disintegration because of lineage rivalries. The main function of marriage, however, appears to be its advantages for lineage recruitment, and our analysis focused on lineage competition rather than group unity as the basis for marriage systems, in particular cross-cousin marriage, which can be seen as a very effective pattern of alliance between competing lineages.

10 *The constraints of culture*

1. Social status

In the previous chapter we made a start on the comparison of human and primate social systems with especial reference to kinship and marriage. The basis laid there can now lead us on to some of the implications for social status. Firstly, let us look again at kinship and status in monkeys.

The first point to note is that in macaque matrilines females compete with one another in a variety of ways. The normal situation is for older females to dominate younger ones, but younger ones can dominate older ones with the help of allies (Datta 1986). What appears to happen is that as she matures, a subadult female assesses the fighting ability of other females, including older ones, around her. She then, at an opportune moment when she has members of her matriline around her, picks on an older, and thus relatively to herself dominant, female of a junior matriline and begins to threaten her. Her matrikin support her and she is able to reverse the rank order between herself and her opponent.

There are also rank order conflicts within the matriline. Datta (1986) reports that in the relationships between rhesus sisters at Madingley, the younger ones, with their mother's support and perhaps also that of middle-ranking sisters, try to outrank the older ones. The oldest sister, however, tries to outrank her mother, and if she succeeds she prevents her younger sisters from outranking her.

In the ranking of matrilines *vis-à-vis* each other, rank depends very largely on fighting ability and numbers. A senior matriline which has been depleted may be challenged and defeated (Samuels *et al.* 1987). It may, however, avoid this by forming a reciprocal alliance with another matriline in which they support each other against a third matriline. This last fact is very interesting in view of the normal antagonism which exists between matrilines, and indicates that monkey lineages can overcome their differences for the greater good of both, which is just what we argued to be the case in human matrilineal wife exchange arrangements (see Chapter 9).

Despite similarities, we need to make a contrast with human status ranking within the family. Between people there are rules of respect

242

where age is concerned, rules of etiquette where food is concerned, and between lineages or families *wealth*, or more exactly the knowledge of relative wealth or power, rather than fighting ability as such is the arbiter of status. Rank remains a vital issue for humans, but there are conventional ways of demonstrating it and actual physical fighting over status is unusual in most societies.

Indeed the whole question of social status is fought out by humans in a different way from that of other species. In many bird species males are adorned with fine feathers whereas females are dowdy. Here, male–male competition is organized around the display of plumage, special posturing and vocalizations, and only occasional fighting. In such species, males who are 'beaten' withdraw from the fray and thus avoid injury and live to compete another day. Some mammals too have evolved display features and here in some cases such as the horns of mountain sheep (Geist 1971) they seem to have evolved not only because of sexual selection but also because of their usefulness in actual fighting. In other mammals, notably some primates such as baboons or gorillas, male–male competition arguably has led to an increase in the size of males relative to females. In gibbons, this size increase has not happened but gibbons in their competition with each other for territory have evolved enormously loud vocalizations, giving rise to what Ellefson (1968) called vocal 'battles' with neighbouring pairs.

The above examples are intended to draw attention to some of the ways in which animals compete. They do fight, but besides fighting they display visual, vocal and postural threats which are recognized as such by fellow members of the species. Competition is over food, mates and territory for the most part. In the case of primates in particular, the competition may be very drawn out, as in the case of male–male competition in chimpanzees (de Waal 1982), and the forms it takes may be very subtle and show the use of a keen intelligence.

In humans, competition from time to time takes on all the forms seen in animals: fighting, self-aggrandizement, possession of territory, though these are culture- rather than species-specific. In nearly all societies, however, there is some kind of wealth: some conventional scarce resource possession of which brings high status. Such wealth may consist of wives and children, or be symbolic and spiritual, such as the cowrie shells prized by the Trobrianders (Malinowski 1922). (See Figure 10.1.)

Wealth of these kinds is not convertible into cash or any kind of generalized medium of exchange, but has high (and carefully calculated) value in the eyes of the peoples who live in the islands between which they circulate. More common by far in modern times is wealth which is

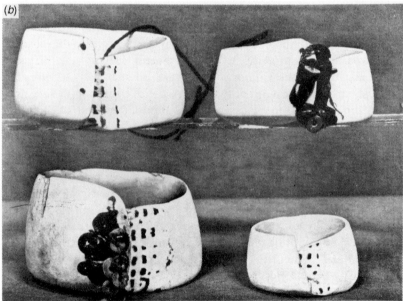

Figure 10.1.   Trobriand armshells (a) Two Trobriand men wearing armshells. (b) Arm-shells. From Malinowski (1922).

convertible into consumer goods or other objects of practical value, and humans the world over at the present time compete for wealth of this kind. Competition may be quite an insignificant part of the lifestyle, as in traditional villages in many parts of Africa or Asia where a cash economy barely exists, or it may be all-pervasive, as in our own society where wealth is seen as the passport to 'success', the latter being in general terms understood as the ability to obtain any desired scarce resources by purchase. Since wealth is a powerful attractant for mates and even marriage partners, will readily buy food, and if there is enough of it will buy territory, then in so far as humans are in competition for these things (and this varies from culture to culture) it is wealth they must strive for. One of the major preoccupations of most human societies is the obtaining of knowledge of the possibilities for obtaining wealth; knowledge of those possibilities is the necessary step before obtaining wealth itself.

2. Wealth, status and marriage

It is necessary to understand the place of wealth in the competition for status in order to understand the ceremonial procedures between lineages at the point of marriage. We noted earlier that whereas in the case of primates the matter of obtaining a mate was a complex business, but not normally a matter for the lineages of the individuals concerned, in the case of humans the families of the pair are normally involved to some extent, and in some cases extensively so. This brings us on to the question of marriage payments, normally in the form of money, goods or labour, or a mixture of these.

What we see, whether brideprice or dowry, is that marriage is a *transaction*. Two families exchange with each other, and in the case of lineages of equal status engaged in bride exchange (see Chapter 9) the flow alternates between them so that neither is the loser. Where lineages are of unequal status, problems arise. For example, if in a dowry-paying society, a low-status family wants to marry its daughter to a high-status family, or (the same thing put differently) a high-status family is involved in the consideration of marriage of its son to a girl from a low-status family, this will involve payment of a large dowry, and this may be difficult or impossible (Srinivas 1962; Dumont 1970). Status considerations are less important in some societies than others. Personal preferences may be paramount. In general, the more the family is involved the more status is important, and vice versa. Perhaps the commonest situation is a compromise, each family gracefully accepting the situation once a marriage has been agreed.

Thus status is a prime consideration in the minds of members of families or lineages engaged in forging marriage links with one another. It is not the *only* consideration. A great many other things may be taken into account, such as health, physical appearance, place of residence, and past history of personal relationships. In primates too, competition for status is not the only form of interaction. For much of the time such competition is absent. Primates spend most of their time feeding, moving around, sleeping, grooming one another, and engaging in exploratory activities relating to their environment. When status competition does occur it can be very subtle, as when one monkey gently displaces another at a favoured spot, or it can be more directly physical. Within a lineage, individual monkeys do from time to time threaten and fight each other, and between lineages they do so too, and blood may be shed. In rhesus and Japanese macaques, males, on reaching maturity peripheralize, emigrate, gain experience (including sexual experience) in other groups, and finally end up in another group where they combine their personal fighting ability with their social skills to establish themselves and achieve status. Females acquire status through their matrilines and from time to time they are called on to defend themselves in agonistic situations. Marriage, in human society, serves as a symbolic demonstration of status, obviating the need for even the occasional reaffirmation of status by hostilities.

This is a second function of marriage. The first, stressed in Chapter 9, was to ensure recruitment into lineages. We now suggest that marriage can serve to disseminate information about the social status of lineages and individuals. This may not always be so. In many human societies clans are differentiated (e.g. by totem) but not ranked. Hunting–gathering societies are usually supposed to be egalitarian in this sense (the North-West Pacific coast native Americans providing an exception, Drucker 1951). Egalitarianism between clans does not of course preclude ranking based on the sex and age of the individuals who constitute the clan. That is a sort of minimal ranking between lineages. In other cases, however, clans and lineages may be quite clearly differentiated by rank, like an elementary class system, as in the case of the Kachin referred to in Chapter 9. In such societies, marriage does nothing to reduce status differences; on the contrary it entrenches and enhances them. In so doing it formulates the shape of the social structure, gives it dynamic expression, exposes it to view, and after the ceremonials are over it leaves its imprint in the minds of all concerned.

In marriage, then, humans have done two new things with kinship – they have developed a way of ensuring continuity of their line, and they have tried to ensure that the *status* of the line is preserved or improved relative to other lines.

3. Marriage and reproductive success

How does marriage relate to reproductive success? The answer to this presents a difficulty. Whereas it is relatively easy to count the number of offspring of a marriage, it is very difficult to be sure that they are all the offspring of both parents. Unless adopted they are very likely to be the offspring of the mother. But paternity is a matter which needs to be established in each case and cannot be taken for granted. A good illustration of this comes from the Marri people of Baluchistan, described by Pehrson (1969). Among the Marri, a desert-living nomadic people, who are by religion Moslems, a woman moves to her husband's camp on marriage. The society is strongly patrilineal, women have few or no rights, and are virtually 'owned' by their husbands. In this culture adultery is taboo, and a wife who is known to be committing adultery can be killed by her husband or one of his agnatic kinsmen. In such a society one might expect that offspring would be biologically descended from their fathers. Such is not, however, likely to be the case at all often.

Marri women do have sexual intercourse with their husbands, but in this society they do not 'love' their husbands, and falling in love is not considered a prerequisite to marriage. Marriage is an arrangement of the husband's and the wife's lineage. After the marriage the wife stays behind at camp when the husband goes off herding his sheep, and during the husband's long absences she is visited by lovers, and he visits his lovers elsewhere. The main thing in life for a Marri man or woman is his or her illicit affairs. These must be kept secret from the public arena. If an affair becomes public knowledge the woman may be killed, and cases of this happening are on record. But as long as men only confide in their menfriends, and women in their womenfriends, all is well. There is a tradition of secrecy about these things. Interestingly for present purposes there is no feeling among men that they must be the *genitor* as well as the *pater* of their children. All children born to a woman belong to the husband and his patrilineage automatically.

Among the Marri the patrilineage is thus *not* a genetic entity. If we assume that each married man has a number of children, but that most of them are not his wife's, then the patrilineage consists of a number of people related socially but not genetically. A man and his sisters and brothers are socially related to their father by very strong agnatic ties, but genetically they are related to their mother's lovers. If we want to trace the passage of the father's genes, we must know all the women who have conceived by him, and we can be sure that they will be far flung, in camps he happened to visit in the course of the movements of his sheep.

In other societies there may be a much closer correlation between the lineage and the flow of paternal genes. Where husbands are able to, and actually do monitor their wives' movements with extreme care, and act jealously whenever the threat of cuckoldry appears, and the society around fosters this, a much closer tie-up between social and biological paternity will occur. It has been suggested (Alexander 1979) that the pattern of lineality reflects the degree of paternity certainty: where human lineages are patrilineal, paternity certainty is high and where matrilineal, it is low, but the example of the Marri is a flagrant exception if that is the case, and probably this theory, derived too directly from comparisons with the mating systems of other species, is oversimplified. We have to accept that the extent to which the pattern of lineality in a human society reflects the pattern of gene transmission varies from very slight to very strong. Probably other factors, such as the influences of religion, contacts with other peoples in the past, and internal processes based on economic and demographic considerations, are the determinants of the pattern of lineality seen in any given society at any particular time.

Returning to the question of reproductive success posed at the outset of this section, we can see from the foregoing that the offspring of a marriage are not always the biological offspring of the parents. The answer must therefore be that the reproductive success of a husband and a wife will be proportional to the extent that marriage brings them into fertile unions with each other or with other partners. This may seem a curious definition but it is precise; it takes into account the possible disjunction between marriage and child production to which we shall return below.

4. The constraints of culture

We have discussed parental controls on choice of marriage partner, especially in relation to status. This feature demonstrated an underlying process: structural constraints on individual choices.

Constraints are everywhere apparent in animal life. The primary constraints are those imposed by the physical environment. Socioecology emphasizes the constraints imposed by society, i.e. by the presence of conspecifics in the local group and their competing designs for available resources.

What we see in humans is a new range of constraints. Just as humans are not free to eat or excrete as and when they please, so they are constrained by considerations of age, sex, social position, and especially role, in all their actions. These constraints are encountered in the socialization process, and we learn to come to terms with them. Socialization is a two-way process. It is not the 'writing on to a blank sheet of a set of cultural instructions'. The individual personality is not a blank sheet but an active component of the final enculturated citizen. What differs between the animal and the human case is the extent of enculturation and, depending on the culture, individuals become more or less amenable and susceptible to parental control of marriage partner. Let us look a little more closely at the process by which conventions, rules, and expectations bring this about, and contrast the situation in non-human primates.

Take the case of a subadult male baboon who had just joined a group in search of a mating partner, as described by Smuts (1985). Primarily he is on the lookout for a female who is not involved in a close relationship with another male, and who responds to his interest. This is not unlike the way constraints operate on young men in search of girlfriends in our own culture, and indeed in many others. We are talking here about mate choice or mate selection, a process involving primarily two individuals in subtle interaction with each other.

Compare the situation when it is not a mate but a marriage partner who is being sought. It may indeed happen that a partner who was originally selected as a potential mate is then later viewed as a potential marriage partner. It is at that moment that, in most societies, family considerations become important. The couple concerned find themselves entering a network of relationships into which they must fit or take the consequences. This network also operates as an insurance after marriage, and as a guardian of marital stability (Bott 1957). Some of the relationships which impinge on a married person are unpleasant. The traditional Chinese mother-in-law had absolute power over her son's wife and frequently reduced her to suicide. Similar cases are known from India and elsewhere. Here we see something not found in non-human society, which profoundly affects the analysis of social process. It is not by virtue of a series of interactions that the Chinese mother-in-law and her daughter-in-law come to grief, it is as a result of the former's expectations that the latter will be totally obedient to her, and the inability of the latter to conform to those expectations.

In such relationships, which we can designate role relationships, interaction is an outcome of the rights, duties, obligations and expectations written into the relationship. And these rights etc. are derived from the social structure itself. The causation of cultural constraints on individual actions can thus be represented as follows:

Social structure ——→ Relationships ——→ Interactions

This, it will be noted, represents social causation in the opposite way to that in which we represented it in Chapter 4, in the discussion of the analysis of non-human primate society. There, following Hinde (1976), we stressed that a meaningful analysis of primate society needed to work from the level of interactions to that of relationships and thence to the level of social structure:

Interactions ——→ Relationships ——→ Social structure

What this seems to indicate is that in human society the information about social situations transmitted by culture can be causally paramount and can determine the behavioural outcomes of individuals. If that is so, we must reconcile this cultural entrainment of individual behaviour with the element of choice and individual autonomy which underlies Darwinian notions of competition between individuals.

The question just put is in some ways the central one for any analysis which tries to argue for continuity of social process from animals to humans. It focuses what has to be explained, namely a novel chain of causation.

We have accepted in previous chapters that the understanding of non-human primate society has to proceed from individual interactions, through relationships, to social structure. We have now seen that the advent of culture and institutions, with roles, obligations and expectations, requires us to think in terms of the reverse direction of causation in humans. We should not oversimplify here. The shaping effect of institutions on the actions of individuals is part of a two-way process, and we need to be aware of feed-back and feed-forward processes at all levels. Nevertheless, the powerful effect of social structures, institutions, role expectations and obligations on the actions and interactions of individuals in human society constitutes a problem for theories of social behaviour rooted in inter-individual competition, in particular if they work outwards from the level of individual competition towards the explanation of higher-level institutions.

What exactly is the problem? Is there not plenty of evidence that competition between individuals is the basis of much of human social life? Some analyses have taken a head-on approach to this issue. Chagnon (1979), Borgerhoff-Mulder (1987a,b) and others have looked at human social situations and have indeed seen an underlying (or even evident, surface) process of competition in human societies. For example, in the Kipsigis, Borgerhoff-Mulder has shown how this competitive process is vigorously pursued by men in competition for wives; some are winners and become polygynists, others are losers and fail to obtain wives at all. Chagnon has demonstrated an even more vigorous competitive struggle among the Yanomamo, who not infrequently resort to violence in their fights, which are very often over infringements of access to wives. Such contests look rather typical of the inter-male competitions over mates seen in a wide variety of animal species, especially polygynous mammals (e.g. lions or Hamadryas baboons).

Are they typical? The question is not so much whether they share an outward similarity with animal contests, as whether the rules of the game are the same. We suggest that, despite similarities, they are not; but we do not therefore abandon the central idea of continuity between animal and human competition.

First, we have already (in Chapter 9) distinguished very clearly between marriage and mating. The correct parallel with animal contests over acquisition or maintenance of mating partners is human contests over acquisition or maintenance of mating partners. There is no doubt that such contests do occur with great frequency in human society, between both married and unmarried adults, bringing about various degrees of disruption to established relationships. Though human mating is hedged about with cultural taboos and restrictions, it seems illogical to deny an underlying similarity with mating generally.

But the competition for wives is something else. It is this that Chagnon and Borgerhoff-Mulder focus on, for a very good reason. It is in the marital relationship that children are born. That is to say, the reproductive success of men and women is most readily (and as a result nearly always) calculated on the basis of numbers of (surviving) offspring, offspring being the sons and daughters of marriage. It is as if a wife, in this analysis, were a kind of 'super-mate' – a mate for whom enormous amounts of effort are expended because of the big reproductive pay-off expected later. Likewise, in societies where female choice plays a large part in the selection of a spouse, a husband can, for the same reasons, be seen as a 'super-mate', i.e. the individual with whom most or all reproduction will take place.

But we have seen that a spouse is something very different from what the term 'super-mate' suggests. A wife is the member of another lineage with whom a husband forms a particular kind of link so that either her children become recruits to his lineage or vice versa. He may or may not pass on his genes through her; as we saw, in the Marri he usually does not.

If that were all marriage was for men, however, we could not expect to find such a high degree of competition. In fact, men and women do marry spouses they find sexually attractive, they do mate with them, and they may do so on an exclusive basis. In other words, there is (analytically) a double-stranded process going on in the human case. There is the socioecological strand:

Competitive mate seeking $\longrightarrow$ mate choice $\longrightarrow$ mating $\longrightarrow$ reproductive success

And there is the cultural strand:

Spouse suitability $\longrightarrow$ betrothal $\longrightarrow$ wedding $\longrightarrow$ incorporation of offspring into lineage

These two strands are in real life inextricably intertwined. But we should note the fact that the cultural strand expresses the corporate will rather than the individual will. Suitability is determined by the community, betrothal is the recognition by the community of the budding relationship, marriage is the solemnization of that relationship, and incorporation of children is their entry into the community's structure, and its rules, roles, rights and obligations. The cultural strand has to do with the incorporation of events into the history and the information base of the society: happenings become social facts.

5. Case studies
The Yanomamo
Let us move from theory to look at the couple of case studies already referred to. Yanomamo village groups live in clearings in the rain forest, separated from other groups by uninhabited terrain. The local group is patrilineal, with a big difference of status between the sexes, men being very dominant. Between men there is considerable rivalry for scarce resources, though kinsmen combine together against men of other lineages. Women are one of the objects of male–male competition, and conflicts over women are one of the causes of fighting between men and between patrilineages. Polygyny is permitted, and one of the ways of achieving high status for a man is to marry more than one wife. As Chagnon (1979) has pointed out, this male–male competition looks very like the competition for mating partners seen in many animal species, and he has concluded that Yanomamo men behave in ways which appear to suggest that they are following (conscious or unconscious) rules to maximize their reproductive success.

Chagnon (1988) has more recently developed his analysis. The only kind of permissible marriage among the Yanomamo is cross-cousin marriage, either matri- or patrilateral, and including first, second or third cousins. Marriage is, somewhat surprisingly in such a competitive society, arranged by the parents of the man if they are still alive. If, as is often the case, they are dead, other relatives will attempt to arrange marriages, and they may favour their own sons, so that conflicts occur. In order to maximize his chances of obtaining a wife or wives, a man or his relatives frequently fudge relationships in order to bring otherwise ineligible girls into the marriageable category. Thus it may be necessary to reclassify the generation in which a particular person is located, for instance by reclassifying a niece as a sister, in order that her daughter becomes marriageable.

Here we have an exceptionally clear example of the informational basis on which marriage decisions are made. Chagnon found that men had a better knowledge of kinship than women (it is the men who do most of the reckoning of marriageability and most of the fudging) as evidenced by the fact that they could name more kin per minute than women could. More pertinently, the number of female kin reclassified as potential wives was significantly greater than expected on a random basis, while the opposite was true for potential wives reclassified as unmarriageable. It further emerged that it was primarily younger women who were reclassified as potential wives, not older ones. Chagnon concludes that in this society at least the question of wife-getting is a major preoccupation for men, that there are rules about eligibility, i.e. there is a set of expectations about who will marry whom, but within this there is an active process of category manipulation to widen the number of potential wives.

This is very interesting in view of what we said above about the corporate processes which can be seen in the choice of marriage partner. Chagnon emphasizes the way individual men spend time and ingenuity re-categorizing women into suitable categories. This shows the importance of the cultural process and its determining role for the actions and choices of individuals. The social structure is paramount: until a woman has been correctly *classified* she cannot be married. The corporate will cannot be disregarded. No animal would have to *re-classify* a female before taking steps to form a mating partnership with her. In human societies which have a lineage structure, it is necessary *because marriage is an inter-lineage institution*, not just something between individuals. This is another demonstration of the difference between marriage and mating. Mistresses and lovers may be taken without any re-classification at all, and indeed in some societies, especially in upper social strata, at various times and places, the relationship of mistress has been reserved for a woman whom it would be unsuitable to take as a wife.

The Kipsigis case

Let us examine next the case of polygyny among the patrilineal Kipsigis of Kenya, studied by Borgerhoff-Mulder (1987a,b). These people combine agriculture with cattle-keeping, living in villages in the African savanna–woodlands. The men in this society are allocated to age sets, marriage is arranged by the parents of prospective spouses, and involves payment of substantial brideprice, paid in cattle. A man's first wife is financed by his father, but subsequent ones he finances himself. Cattle, land and cash are major preoccupations of men, who differ from each other greatly in ownership of these scarce commodities. The parents of girls involved in marriage transactions are primarily motivated by considerations of the man's status, and that of his family, and they attempt to secure the best possible marriage for their daughters.

Borgerhoff-Mulder found a statistically significant positive correlation between a man's wealth and the number of his wives and offspring. While some poor men cannot afford to marry at all and many have to content themselves with one wife, others marry more often, and one man had twelve wives. In the Kipsigis marital fidelity is high; polygyny is the major cause of the differentials in reproductive success between men. To support many wives a man needs much land, he can acquire land through the possession of cattle and cash, his immediate strategies thus revolve around increasing his holdings of these physical things, but possession of these is not an end in itself, it is the means to wife acquisition and the production of heirs.

Once again, in analysing the Kipsigis case, we can see the corporate process at work. It is not a man himself but his father who pays the bridewealth for his first (perhaps his only) wife. The woman's parents are primarily concerned about the man's status; they will receive the bride-wealth and they will withhold their daughter from a man they consider unable to pay. Lineage controls are apparent everywhere.

Paternity

Regarding the issue of paternity, the Yanomamo and Kipsigis cases are more representative of traditional human societies the world over than are the Marri. Borgerhoff-Mulder (1987a,b) states that adultery is almost non-existent among Kipsigis. Chagnon likewise emphasizes the serious-ness of adultery and the legitimate revenge of a man cuckolded. So it is in very many societies where life is lived in villages, the world over. The Marri are nomadic, and it is probable that adultery is more common where mobility is greater; it may be tacitly tolerated to a greater or lesser degree, or lead to a high frequency of marital break-up, as among the hunting–gathering Hadza (Woodburn 1968). In city life, again, adultery seems to be commonplace, the anonymity provided by large numbers enabling such affairs to remain hidden from the married partner.

The production of offspring outside of marriage is handled quite differently in different societies. Nevertheless, much or even most mating which leads to children is marital mating. Marriage is essentially an institution which provides a culturally suitable and legitimate context for reproduction, and this is entirely in keeping with the idea advanced earlier that it is concerned with lineage recruitment. If it appears that we use the Marri example when it suits us and not when it doesn't, this is not in fact the case. We used the Marri example to show the conceptual and analytic distinctness of genetic and social transmission, which it does well. But the Marri example cannot serve to generalize about the actual frequency with which genetic and social transmission are identical or divorced from each other. It is but one example among many, and in fact lies at one end of a continuum, the other end being cases like the Kipsigis where almost all mating is marital mating.

6. Evolution of marriage

In the present analysis we have tried to move between biological and sociological frames of reference and to find links between the two. This provides a new method for approaching the old questions of how and why marriage evolved. Levi-Strauss (1969) saw marriage as a solution to the problems of incest and a way of bringing about social solidarity. We have seen that it can be of mutual reproductive benefit to lineages to engage in wife exchange. We have also seen that systems of kinship and affinity are systems of knowledge – knowledge about rights and expectations and obligations between those who know themselves to be in particular positions relative to each other in the structure of kinship and affinity. How might such a situation have evolved?

The answer must involve an intelligence of a particular kind. It cannot just be a high degree of general intelligence, for this puts intelligence on a linear scale which does not do justice to the variety of ways in which it can express itself. Rather, in order to build a human social system, our ancestors needed an intelligence capable of inventing new categories, and then of labelling and reifying them. We know that primates can distinguish between 'own lineage' and 'other lineage' (Datta 1986), or 'mother–infant' and 'not mother–infant' (Dasser 1988), and we know also that monkeys can categorize objects such as leopards, eagles and snakes, and distinguish between them in vocal communication (Seyfarth *et al.* 1980a,b). Equally important, or maybe more important from the culture-as-information-processing standpoint, is that they appear to use these categories in communicating with one another.

The human development in which we are interested is a very particular extension from what was already possible. Social thinking of a high order was already in existence prior to the invention of marriage. Our pre-human ancestors would have been engaging in tactical deception (Whiten & Byrne 1988a,b), alliance formation (Packer 1977), altruistic interference on behalf of kin (see Chapter 9), reconciliation after quarrels (de Waal & van Roosmalen 1979) and a whole variety of other complex social interactions based on skills and intelligence. They would already have had the ability to devise categories and communicate them to others. Cultural transmission of object manipulation, such as food cleaning or tool use, would be present. Lineages existed as distinct biological entities, each having a particular relation to the others.

The further emergence and recognition of lineages as distinct cultural entities depended on the selective advantages which arose from the establishment of particular relationships between lineages. Such an advantage, we argue, arose because there were reproductive benefits to be gained, namely recruitment benefits. Thus, instead of each individual seeking mates competitively, arrangements were made within and between lineages for the provision of mates according to newly invented, logical rules. In this the older generation would be influential in directing offspring towards appropriate mates, leading eventually to lineage exogamy.

From these early beginnings have arisen all the different social systems in existence today. Social anthropologists have shown how some kin are classified together while others are clearly distinguished, and one of the central objectives of social anthropology has been to discover the different kinds of logic by which such different schemes become 'normal' to their practitioners. Kinship is a matter of logic in the human case. The logic differs from society to society. For instance, most traditional people in Africa or Asia would be very surprised to learn that we in our society designate a father's brother by the same term as a mother's brother. The reason we can use the term 'uncle' for both is that in our bilateral kinship system the expectations, rights and obligations of uncles towards their nephews and nieces are the same on both the father's and the mother's side. Such would not be the case, for example, in a patrilineal society such as the pastoral Nuer (Evans-Pritchard 1940), where a father's brother has rights and obligations similar to those of a father, while a mother's brother does not, and is quite unimportant in the lives of his sister's sons and daughters. By contrast, in the matrilineal Trobriands (Malinowski

1922) the mother's brother is the most significant man in the life of his sister's children, for he has powers of authority over them, while the father's brother, belonging as he does to a different lineage and clan, has no such authority, though he has other rights.

Thus we see that all human societies have become divided into subgroups which exist in opposition to each other and have particular relationships with each other. This process of differentiation has a logical basis in each case, and each case has its own logic. The logic of a particular instance is seen by which kin are terminologically equivalent, and which are differentiated. Mother and father are always differentiated, but their same-sex siblings are often terminologically equivalent, so there is an element of inclusiveness in kin categories in respect of the relationships they describe, and indeed the kin terms we find are based on inclusive logic, rather than spelling out in precise detail the exact relationship between individuals. We should not expect absolute clarity in kinship systems, though we may strive for clarity in describing them. We have already seen in the Yanomamo case that individuals may find it in their interests to muddle up kin relationships, so that more or less can be made of the significance of particular relationships according to the personal ambitions of the participants in a system.

7. Advantages of social networks

Gamble (1980, 1982) has written interestingly on the interactions between social groups in palaeolithic society. He draws attention to the function of display items such as figurines in the exchange of information between groups. Noting the common stylistic elements of such art objects (and, e.g. cave paintings) over large areas of Eastern and Western Europe, he points out that a common style 'stems from common encoding and decoding strategies within a system of visual communication'. He adds that without a common conceptual framework, there can be no communication (1982, p.99). Following Wobst (1977), he concurs that sending and decoding messages

> is especially useful when contact is being maintained with a socially distant population. Interaction with such a group may be infrequent, with potential sources of error arising from language differentiation. Visual communication may be sufficient to sanction and direct behaviour at moments of contact and thereby make social intercourse more predictable when it does take place.

Wobst (1976) proposed the analytical concept of a 'mating network' (perhaps roughly equivalent to the social anthropologist's concept of the

Figure 10.2.   Sculptures and engravings from the Upper Palaeolithic. From Jolly & Plog (1979). Courtesy of the American Museum of Natural History.

endogamous tribe) which 'defines the largest social entity beyond the local group which has behavioural concomitants' (1976, p.49). Gamble (1982) discusses the mating network model, and writes:

> The mating network model therefore provides a picture of the arrange-
> ment and location of hunter–gatherer personnel, from which we might
> make inferences about marriage patterns. For it to function there must
> be several rules. Exogamy forms the most common marriage rule
> between members of local groups, while further rules also exist concern-
> ing access to food and other resources . . . A complex pattern of
> alliances is therefore created in order to distribute personnel over an

area. These alliances take several forms. They include ties established through marriage . . . These alliances, and the rule systems of reciprocity that maintain them, are means by which social and biological reproduction are achieved. (1982:100).

And further:

> I would suggest that when we talk about interaction and information exchange in the palaeolithic, we are in fact describing the evolution of this alliance system . . . the difference for example between neanderthal and cro-magnon populations and their distinctive material culture was not derived from differences in biology or intelligence. Rather it stemmed from differences in their societies, the structure of their alliances, that resulted in the potential information content of material culture being used in alternative ways.

The great advantage of inter-group alliances would be that:

> local groups would be bound together in particular mating networks even though the constituent member groups were not necessarily the closest in spatial terms. The proximate groups that lay outside the mating network may however, have been important in times of resource shortage when access to alternative food supplies was required.
>
> This model therefore recognizes two alliances. The first set concerns membership within a mating network where in effect a closed system comes into operation beyond a certain spatial scale. The second set of alliances establishes a framework for coping with expected shortfalls in resources. A local group would have a series of overlapping alliances formed for different purposes. On the one hand the process of boundary maintenance, important for the mating network, would be enhanced, while on the other networks of alliances would have extended beyond such a system for the simple reason that factors of distance and the distribution of resources prevented the mating network from being self-sufficient.

Gamble here pushes the idea of alliances between palaeolithic groups even further than we have done. We have stressed the advantages to individuals and lineages of strong marriage links with other such groups in terms of social recruitment. Gamble emphasizes not only the mating structure but also the question of fluctuations in food supply, and sees the exchange of artifacts as a long–distance method of communication.

8. Language

Primate kinship systems are based on evolutionary logic, the logic of survival and reproductive success. In human kinship systems, that logic is transformed by the introduction of symbolic language, with a new emphasis on contractually defined rights and obligations. The logic of human systems keeps in existence sets of verbal regulations which have the effect of providing appropriate marriage partners, and of keeping wealth within the lineage. Human kinship systems are, we suggest, logical systems of thought and action expressible in linguistic form. They are based on the underlying logic of living systems generally, namely the logic of the evolutionary process, which endlessly promotes any kind of action or behaviour which improves reproductive success. In the final chapter we shall look at the question of the evolution of language, and try to relate this to the evolution of human systems of kinship and affinity.

9. Conclusions

In this chapter we have maintained the focus on kinship and marriage established in Chapter 9, but have moved on from comparison with non-human primate kinship to a consideration of how human systems of kinship and affinity can be understood primarily as extensions of the socioecological constraints we have encountered throughout this book. To recapitulate, those constraints are of two kinds, environmental and social. Our analysis has focused on the latter kind, and we have stressed the role of information transmission and processing as the way primates deal with these constraints. In the present chapter we saw that human cultural institutions exercise a decisive controlling influence on the actions of individuals, and can thus be understood in the same terms as we have used for the understanding of non-human primate social constraints.

A paramount constraint in human cultures is the pursuit of symbolic wealth, taking different forms in different places. Competition is channelled into wealth-seeking, and access to scarce resources is dependent on success in the conventional competition. Such scarce resources include not only material objects but also social ones, notably in the competition between men for wives, and we show examples of how this competition works itself out in different societies. More generally, it seems that lineages are in competition, as they are in non-human primates, and that they seek, in their marriage alliances, to maintain or improve their status. This explains the extent of interest in such events by all members of the kin networks involved. The extent to which marriages are motivated by, and lead to, reproductive success is a problematic area and varies between different societies. It is shown that we need to be very clear-minded about the difference between mating and marriage, and to investigate the relationship between the two in particular cases.

The rules holding kinship systems and systems of affinity together are logical rules, but the logic differs from society to society. Where the rules impede the formation of marriage alliances they are frequently bent in individual cases. Here we see the manipulation of information by those who can – the powerful. Human institutions are man-made, and their supporting rules are capable of being re-interpreted as a result of discussion. Indeed the edifices of human cultures could not have been constructed without the use of symbolic language, enabling the designation of roles, to which the formulation of appropriate rules of action could be applied. Marriage evolved as a powerful structural linking mechanism, and undoubtedly had advantages in the bringing about of extensions of the communicative network between palaeolithic groups.

11 *Language and its social implications*

1. Introduction

In earlier chapters we stressed the cognitive nature of social interaction in primates and the fact that their social structures are made up of relation-ships based on learned information about other group members. We have come to see the interactions between primates as involving both compe-tition for and cooperative investment in local resources, with cooperation typically characterized by series of transactions in which benefits of one kind are exchanged for benefits of other kinds. We have avoided the grosser kinds of analysis based on close adherence to genetic affiliation or simplistic notions of selfishness in the explanation of social behaviour.

In Chapter 9 we compared kinship and mating in non-human primates with kinship, mating, and marriage in humans, and saw an underlying level of commonality with a series of cultural extensions in the human case. In Chapter 10 we emphasized the dimension of social status in the organization of kinship and the marriage process, and saw how this preoccupation with status was found in both non-human and human primates.

Demonstration of parallels in the social life of contemporary monkeys, apes and humans is interesting and suggestive of evolutionary continuity, but of course it does not constitute an explanation. We have shown that there are similarities in the organization of social action and of societies of hominoid primates, but we have not dwelt on how those similarities might have come about. This was no accident; we did so to avoid speculative arguments of the 'if I were a horse' variety (Radcliffe-Brown 1952).

Such avoidance of speculation is praiseworthy if not taken to extremes. Anthropologists who have managed perfectly to subdue their imagin-ation make dull company. Only informed speculation can give us a sense of how our society evolved and what possibilities lie ahead. However we may theorize about evolution, it is realized in impinging series of contingent historical events (Gould 1989), events of particular character in particular order. Clearly it matters how we conceive of, for instance, hunting as practised by our Pliocene or Pleistocene ancestors, not just for getting the prehistory texts right, with reasonable representations and reliable chronologies, but for shaping research in palaeoanthropology. As can be seen in retrospect, those who argue that a hunting way of life was critical to human evolution must be clear whether by hunting they have in mind meat procurement by means similar to those practised by contemporary and near contemporary peoples or something more akin to the hunting/scavenging behaviour of certain large mammalian predators. If the former, then we may look for evidence of a dual economy rooted in a primarily sexual division of the labour involved in procurement and extraction procedures appropriate to animal versus plant food species, and shifting home bases or less well-defined central locations (Isaac 1978a,b, 1983) indicative of food transport, temporary storage, and redistribution – i.e., of *reciprocal* sharing in a *dual* subsistence economy. If the latter, with no need to postulate a dual economy (as seems more likely through at least the Pliocene and Lower Pleistocene), then different evidentiary considerations claim palaeoanthropological research attention.

Alternative scenarios abound concerning the origin of hominid culture between 5 and 3 mya in conjunction with male provisioning (Lovejoy 1981), hunting (Washburn & Lancaster 1968), gathering (Tanner 1981, Zihlman 1978, 1981), food exchange (Isaac 1978a,b), scavenging (Shipman 1984), hunting–scavenging or scavenging–hunting (Geist 1978; Bunn & Kroll 1986), or yet some other mode of subsistence behaviour, each scenario involving speculative reconstruction of some complex adaptive transformation of primate anatomy and behaviour in response to environmental change. Such scenarios may derive primarily from archaeology (Isaac 1978a), primatology (Wrangham 1980), or ethnology (Tanner 1981), but tend in any case to have multidisciplinary referents; increasingly the attempt has been to describe proximate constraints and socioecological relationships in detail, using evidence derived from taphonomy, palaeobotany, palaeodemography, etc. (e.g. Brain 1981; Foley 1987, 1988; Standen & Foley 1989a).

An approach that primate behaviourists have employed to good effect recently in comparing primate social behaviour across species, and across groups of species, but which has not hitherto been adapted to understanding the evolution of society *per se* is the study of cognition, and especially of social cognition. The cognitive approach does not emphasize holistic scenarios of ancestral hominid behaviour so much as analysis of *what changes in information processing may have been entailed* in moving from the kinds of society available to our pre-human ancestors to the kinds of human society seen today. As our actual pre-human ancestors are not available for inspection, we must turn to socioecological considerations of primate cognition in action, not of course to particular species as analogue models – e.g., baboons or chimpanzees (or bonobos) as most closely representing pre-hominids in this or that respect (see criticisms of this approach in Kinzey 1987) – but to comparative evaluation of those particular aspects of cognition which we see demonstrably underlying the social life of all non-human primates and which has formed the subject of this book in chapters 2–8. This approach leads to construction of a *processual* model (Tooby & DeVore 1987), a model of the way in which the processes underlying primate action changed during human evolution to produce the very special kind of society found in humans alone.

Our discussion of primate social behaviour has emphasized the way in which monkeys and apes handle information of socially relevant kinds. Socially relevant information includes information about status (of one-self *vis-à-vis* another, or of others *vis-à-vis* one another), about sexual condition, about pairedness (whether another animal has a possessive mate or not), about kinship. All this information (knowledge underlying individual action) is carried in the minds of primates living in society and is brought to bear by them on the making of decisions from moment to moment during everyday life in the group. The same, exactly, can be said of human beings. The difference between human beings and other primates lies primarily in human beings' use of language for processing information.

2. Language, social cognition and culture

Recall that in Chapter 6 we defined *social cognition* as the application of intelligence to the management of social knowledge (i.e., information concerning the identity of conspecific others, the character of their behaviour as individuals, and the specific nature of their relationships with one another, including oneself) in conjunction with the exploitation of social relationships. Life in society imposes constraints on individual cognition and individual action in competition for local resources; at the same time, individuals in groups are afforded advantages in procuring and utilizing resources – e.g., as individual foragers competing and cooperating with one another – by the wider social processing of environmental information (information which is perceptually decoded as opposed to information which is genetically transmitted) which life in society affords.

Culture, it seems to us, when looked at from an evolutionary stand-point, is most usefully treated in processual terms, and specifically in terms of those constraints on and advantages to individual cognition and action just noted. This view of culture is at once broad, applicable to every group-living animal species, and sharply focused on cognition and social interaction. It will not suit every anthropologist, but its inclusiveness makes it appropriate for those comparisons across species which are essential to studies of evolution, and the sharpness of focus helps ensure that comparison will prove productive.

We observed in Chapter 8 that there are three repositories of cognitively processed environmental information: individual memory, working memory collective to a group, and artifactual remains of behaviour. The kinds and amount of information filed in these repositories differs from species to species; so, apparently, do particular features of storage, retrieval, transformation, and social transmission of information, even in clades in which more general commonalities derive from a common structure underlying perception and cognition (e.g., primates or for that matter mammals). At present, quantitative comparison of differences of either sort are extremely rough (e.g., endocast or cranial vault measurements as the basis for comparisons of hominid storage capacity; presence or absence of Broca's area, inferred from endocast structure or endocranial impressions, as an index to character of processing). Our very tentative suggestions in Chapter 8 concerning differences in information processing for monkeys, apes and man were offered as an indication of what needs to be compared, not as a preface to quantitative comparison. The primary utility, at this date, of categorizing information repositories and noting that each category presents different problems with respect to processing – and especially to *social* processing of information stored therein – lies in what even such a rough gesture toward analysis may suggest concerning the relation between culture and language. What it suggests to us is that *more is to be learned from asking how language may be rooted in culture than by asserting that human language is the difference which gives rise to culture.*

We will not speculate on the precise circumstances of language origins – on the other hand, it is important to distinguish between evolutionist and saltative or essentialist accounts of *glottogenesis*, the emergence of vocal language – and in the sections to follow we will look at what each has to contribute to our comparison of human versus non-human primate information processing. Our hominid ancestry can be traced in the Early Pleistocene to a radiation of forms which occurred in conjunction with a

Plio-Pleistocene econiche transformation. That transformation evidently involved a shift to terrestrial bipedalism and associated changes in diet and foraging practices for the hominines concerned. While paleoanthropologists continue to argue about how many species of hominids may have coexisted in either South or East Africa, at any given evolutionary moment of the Plio-Pleistocene, and how relationships between or among those species ought to be characterized, there is wide agreement that from around 3 mya until not quite 1 mya *at least* two species, one of gracile, one of robust form, must have coexisted. One or perhaps both mastered tool-making (Susman 1989), but apparently only one evolved a cognitive capacity for language measurable in an expansion of that portion of the brain in which Broca's area is located (Falk 1987). Whether these species ever occupied the same habitats simultaneously, or, if they did, whether groups in sympatric populations ever competed directly for identical resources is impossible to say solely on the basis of evidence from archaeology, palaeontology, and comparative anatomy. However, the disappearance of the robust form shortly after the appearance of a new (Acheulean) toolmaking tradition, and in conjunction with gradually increasing brain size for *H. habilis* and *H. erectus* (insofar as the relatively few crania available permit characterization of an evolutionary trend in brain size through the Early Middle Pleistocene), suggests that evolution of language and culture, however these may have proceeded, probably resulted in an increase in numbers and, arguably, niche expansion for the gracile form, consequent increase in the occasions and intensity of direct interspecific competition, and eventual extinction of the robust australopithecines.

3. Stages of linguistic culture[1]
Gordon Hewes has suggested, drawing on evidence from a number of disciplines (Hewes 1973a,b, 1976, 1978, 1983, 1989) that language evolution hinged initially on postural/gestural, not vocal components of communication. Early gestural language presumably would have been *sememic*, with units of meaning carried by morphemes. We would suggest that it must have been so by definition, if it did not seem unreasonable somehow to insist that analytic categories which we use in talking about language today must apply with equal propriety to discussion of a proposed predecessor system – but note that there is nothing about the definition of a *sememe* which restricts its application to constituent features of a strictly vocal language. From the start, though more likely at some later stage in its evolution, gestural language may have been vocal

in the sense of being not just *sememic* but *phememic*, containing vocal signs of distinct semantic as well as phonological character – i.e., *phememes*, as opposed to *phonemes*, in which semanticity is greatly diminished or absent (Foster 1983; Hewes 1983). In other words, gestural language could have accrued and probably did accrue vocal as well as gestural signs and perhaps vocal qualifiers, themselves sememic, over- lying gesture (Milo & Quiatt in press b). However, in Hewes's view, glottogenesis proper would have entailed phonemicization of language, a transformative event contingent on adaptations, biological and cultural, within a context of selection which only long-accustomed use of language on a species-wide basis could provide. The two-phase aspect of Hewes's gestural theory of language seems to fit the facts of hominid evolution, in its anatomical aspects, and, in so far as they can be inferred, of major shifts in Pleistocene hominid subsistence and foraging patterns. After the disappearance, at around 1.5 mya, of robust australopithecines from the fossil record, hominid evolution appears have been anagenetic in charac- ter and conservative, in terms of cultural as well as anatomical change, for almost a million years. During that time, according to Leigh (1988) we find a steady but gradual increase in brain size, of the order of 50 cc. per 100 000 years. Then cranial capacity increased more rapidly, 200 cc or more to 1350–1600 cc for so-called archaic and sapient remains from the Late Middle and Early Upper Pleistocene. In addition to this increase in brain size, qualitative differences in features of the post-cranial skeleton and the bimodal character of quantitative differences in others suggest to some (e.g., Stringer 1984; Stringer & Andrews 1988) cladogenetic evolution, selection and adaptation of two species to different circum- stances, and, in Western Europe at least, eventual direct competition culminating in the disappearance of one species from the fossil record. The situation is reminiscent of that in the Plio-Pleistocene. The main difference is that in this latter period events are speeded up, less than 200 000 years elapsing between the appearance and the extinction of Neanderthals (Zubrow 1989). In this admittedly controversial interpre- tation, the outcome of interspecific competition, as perhaps in the earlier instance of competition between gracile and robust hominids of the Lower Pleistocene, would have been partly or largely contingent on differential achievement of language capabilities (Milo & Quiatt in press a,b).

The advantages of a fully vocal language, syntactical and phonemi- cized, over a language consisting simply or primarily of a vocabulary of gestural/vocal sememes seem obvious. Most obviously, the vocal–

auditory channel in itself affords broadcast communication and multi-directional reception by day and night, largely independent of habitat and climatic variables (Hockett & Altmann 1968). Syntactic organization of utterances increases precision in expression of meaning without sacrificing either flexibility of statement or economy of utterance; well developed syntax seems particularly important in connection with statements about subtle or complex relationships between entities, the organization of spatial fields, and mode of action – all of which would have been important to hominids who were proficient collectors of plant and animal food species, foraging out of temporary base camps, with regular division of labour by sex in a dual reciprocal subsistence economy. That groups of australopithecines or of *H. habilis* were organized at this level of socio-economic complexity around a hunter–gatherer subsistence seems unlikely (Binford 1981; Brain 1981; Potts 1983; Shipman 1983, 1984 – but see Tooby & DeVore 1987.) However, most interpretations of the more extensive faunal and tool assemblages recovered from late *H. erectus* sites in Europe and Asia are congruent with the assumption that such an economy was in place, as a socio-cultural feature of that species' niche, toward the close of the Middle Pleistocene.

The advantage of phonemicized speech over sememic but non-phonemic vocalizations may have had as much to do with economy in the storage and retrieval of information as with communication *per se* (Hewes 1983). Hewes argues that phonemicization would have established the basis for organizing words and phrases in memory by initial sound, in something like an alphabetized file, prior to the invention of alphabetic writing. Also, and similarly, phonemicization in conjunction with syntactic structure and a rhythmic line would have enabled a mnemonic system of end-rhymes, front-rhymes, and internal sound correspondences of many sorts, contributing further to memorization of rituals, recipes, formulae, and longer recitatives central to the cultural life of hominids. Such a system would be useful not only for individual storage and retrieval of organized sets of essential information ('Thirty days hath September . . .') but, to return to speech communication, for transmitting such information within and across generations. Above all, perhaps, it would facilitate synchronized retrieval, iteration, reworking, and restorage of information sets, by several or all members of a group interacting in formal and informal contexts, as a way of affirming group identity and maintaining the common cultural store of information (c.f. our Appelachian scenario in Chapter 8).

Syntax itself contributes to organization of words (as parts of speech) for efficient storage and retrieval according to function; the conjunction of syntax and phonemicization must have afforded considerable increase in the size of a language community's working vocabulary. Acheulean hunters/foragers may not have required vocabularies larger than a few hundred words. If not, achievement of language performance intelligible by fellow members of the community should have been simple enough. *H. erectus*'s culture may not have varied greatly across wide regions, judging from the relative homogeneity of the central tool-manufacturing tradition, and, assuming that language at that time was primarily gestural in form, component units may well have been tied rather closely to a narrow range of subsistence activities and to technological processes immediately recognizable throughout the species. Thus, for *Homo erectus*, translation and learning of language across communities, however widely separated geographically, may have been akin to picking up a dialect.

In the Late Pleistocene, artifacts and bone assemblages evidence greatly increased proficiency in hunting and, to varying degrees according to habitat, concentration on a limited number of preferred species of prey (Gamble 1986). In some regions (e.g., Western Europe around 40 000 years ago) this apparently led to increased population density, territorial restrictions on group migration, and more intensive exploitation of local resources, i.e., a counter-tendency towards an increase in the number of food species utilized. Ramification of tool types, local diversification of form, and stylistic elaboration in the Middle and Upper Palaeolithic evidence increasing heterogeneity of information arrays relating to basic subsistence, not just across but within geographic regions. They also suggest an increase by an order of magnitude in the store of information appropriate to the distributed diet and scheduled collecting of foods necessary for groups who must, in the face of increasing population density, engage in more intensive exploitation of an expanded list of food species. If the language of *H. erectus* was, as has been suggested, gestural at base and sememic in character, it would have been hard-pressed to serve communities in these new circumstances. Gradual increase in the brain size of *H. erectus* may reflect selection for greater efficiency of information processing in a species in which cultural information is organized in a sememic system, the 'natural' boundaries of which (see above) are highly limited. The jump in brain size in the Late Pleistocene suggests rapid marked change in the intensity of that selection. Strong selection, in conjunction with temporary isolation of some populations and reduced migration overall, after the first dissemination of *H. erectus* throughout the Old World, could well have resulted in one or several avenues of language development. Any innovation which provided a more conventional, less limited system of organizing information for storage and processing would have been advantageous.

The characteristic form of (classic) neanderthal crania, frequently accounted for in terms of adaptation to cold or other external environmental condition, remains unexplained. It may be that neanderthal and sapient brains of the Late Pleistocene, which are of similar size but possibly of different architecture, represent alternative responses to selection for faster processing of more information (Milo & Quiatt in press b); whether they do or not, it is interesting to note that the increase in both cases is achieved well in advance of the efflorescence of Upper Palaeolithic culture, with its familiar graphic and symbolic components – by which time most would agree that a phonemic language is in place and characteristic of *H. sapiens* as a species. Phonemicization, it should be noted, could have been followed by relaxation of selection for increasing brain size – if, for example, the efficiency of retrieval made possible by phonemicized speech, along with, in practical terms, infinite productivity, made further increase in size irrelevant.

Early hominid fossils and tools yield very little evidence concerning the origins of language, and what there is is indirect. Falk's comparisons (e.g., 1987) of Australopithecine and *H. habilis* brains suggest that the latter may have been undergoing neural reorganization, developing the cognitive bases of language of some sort, but her conclusions are suggestive at best – and suggest nothing in any event about linguistic *expression*, gestural, vocal, or other. It is of course important to distinguish between language and linguistic expression, for language *per se* 'is not a type of sensorimotor activity in the mouth and ears or the hands and eyes, but a cognitive process in the cerebral cortex' (Wescott 1974). Where expression is concerned, it is reasonable to assume that glottogenesis occurred in conjunction with evolution of 'the peripheral organs of speech (larynx, pharynx, palate, tongue, etc.)' (de Grolier 1989; see also Lieberman 1985); the key event arguably being that descent of the larynx which is observable in human ontogeny but not in that of other primates or of mammals more generally. Lieberman since 1968 (e.g., 1976, 1983, 1984, 1985, Lieberman *et al.* 1972) has promulgated a controversial theory which postulates highly specific dependent relations between the architecture of the human vocal tract and features of oral language essential to efficient transmission of information. These relations he maintains, on the basis of perhaps too imaginative reconstruction of the Neandertal vocal tract, could not have obtained in the case of Neanderthal and may have resulted in the extinction of Neanderthal populations in competition with anatomically modern *Homo sapiens*. Laitman (1983,

1985; Laitman *et al.* 1989) makes an argument which is similar in its emphasis on late evolution of the anatomical bases of human speech but is less extreme and more persuasive, being founded on careful, extensive anatomical comparisons across mammalian species. Whether or not descent of the larynx (as a phylogenetic incident) occurred when Lieberman asserts that it did, or whether it is crucial to assessment of the vocal language capabilities of hominids, as both Lieberman and Laitman say it is, remains in dispute (see, e.g., de Grolier 1989; Wind 1989). Laitman's observations fit our own notions concerning the late development of vocal language; but, again, one can only speculate as to the final implications for speech of observed anatomical differences.

Still less can inferences about language evolution be drawn directly from analysis of artifactual remains in the Early and Middle Pleistocene. There may be, as Hewes has argued, a close connection between object manipulation and tool manufacture on the one hand and, on the other hand, the employment of gestures for communication (see also P. C. Reynolds 1981); there may be, as Hewes also has argued (and no one has made more thorough enquiry into these matters), persuasive logical reasons as well for postulating a prolonged stage of gestural language prior to glottogenesis. However, the artifactual remains of hominid culture, which through the Middle Pleistocene are comprised almost wholly of flaked stone, provide precious little information about the character of subsistence and social organization, let alone language in either its cognitive or its expressive aspects (Wynn 1979, 1981, 1985, and especially 1989a,b). What we are presented with is a technology which apparently did not require a high order of intelligence to manufacture, in either its Oldowan or Acheulean manifestations (Wynn 1979, 1981)[2].

Perhaps the strongest indirect evidence for language use (and we have noted that there is nothing *but* indirect evidence through most of the Pleistocene) is regional diversity of culture in the later Pleistocene, in which material remains are both abundant and highly differentiated in terms of inferred function, and in which discontinuity of patterns from one area, in the absence of significant geographic barriers, argues the exercise in unprecedented strength of sociocultural control over the flow of genes and ideas. There is of course a problem here, given the adverse effects of time on material remains of complex and relatively insubstantial structure – the longer the span between the time things are used and discarded and the time archaeologists set about their work, the less likely it is that their existence will be remarked. Hence it is not surprising that in interpreting the lifeways of hominids two million or a million years ago we have little to go on but stone and the marks made by stone on bone, and that as we move closer to the present we begin to find artifacts of other materials and of greater variety. We should not jump to conclusions about relative complexity of cultural and informational processes from either the paucity of the earlier remains or the abundance of the later. It is, to repeat, the *regional diversity* of remains in the Late Middle and Upper Pleistocene, not just their relative richness (in terms of abundance and degree of functional differentiation) which argues a new level of efficiency in the processing of information in conjunction with (and probably responsible for) increasing cultural barriers to inter-regional communication in a still rapidly evolving species (Milo & Quiatt in press a,b).

Ultimately, perhaps, resolution of the problem of how and when phonemicization of language occurred will depend on construction and refinement of theory. We should not anticipate learning much directly from archaeology, palaeontology, molecular genetics (nuclear or cytoplasmic), or comparative anatomy; however dramatic the impact of yesterday's or today's discovery, the new data provided rarely tell us much in and of themselves. Continual comparison and testing of theory against all available data will gradually clarify interpretation of the character and significance of what evidence there is and will assist in correctly assimilating new evidence as it comes to view. In this process an evolutionist stance with regard to language origins seems to us the posture of strength. An essentialist position appears to offer little help toward understanding how biological and cultural evolution are related. However, Noam Chomsky's essentialist theory of human linguistic behaviour (see, e.g., Chomsky 1967, 1970) has been useful in clarifying the general structure of language and raising questions about its character as a species attribute, and so we turn briefly to consideration of Chomsky's ideas in the next section.

4. Language acquisition by children

The evolution of spoken language is commonly perceived to be humankind's distinctive behavioural 'achievement', and we share this perception. It may be difficult to conceive exactly how an evolutionary increase in brain size from, say, 1150 cc to 1350 cc may have advanced social or for that matter individual processing of information, apart from a presumed increase in storage capacity, but it is quite clear what most of the implications are where the development of spoken language is concerned. We have discussed many of these, and we summarize them for review in Table 11.1.

Table 11.1. *Informational features of vocal language compared with gestural language*

	Vocal	Gestural
Features relating primarily to transmission of signals	Broadcast transmission Omni-directional reception Signal independent of light source Rapid fading Volume control Transmission channel open in normal waking state Signal generation relatively independent of other activities. (Drinking and swallowing food are important exceptions.)	Directional transmission Position-dependent reception Light-dependent signal Fading subject to control Signal strength relatively constant Non-linguistic cue may be necessary to alert receiver Signal generation may be interrupted by or incompatible with activities such as locomotion, object manipulation, predator defence
Features relating primarily to intra-individual storage and retrieval of information	Clear distinction between -etic and -emic features facilitates systematic coding and filing (e.g., alphabetized filing, or the phonetic equivalent thereof).	?Presumably less important. However, where general system features are concerned it seems reasonable to withhold confidence from assessments made by someone habituated primarily to one or the other of the systems being compared – hence the question mark.
	Formal similarities of certain -emic structures may provide important mnemonic cues (as in rhyme schemes, e.g., 'Thirty days hath September . . .')	?Presumably less important.
Features relating to overall system operation	Correspondences and contrasts between both -etic and -emic features of signs provide a basis for contribution to instruction and to generation of problem solution via witty juxtaposition of images, punning, etc.	?Probably about the same
	Approximate analogues to vocal signals of sounds in nature, and context-dependent mis-interpretation of other sounds (or auditory hallucinations) provide a basis for revelation as a form of self-instruction and stimulus to behaviour modification.	?Presumably less important – auditory hallucinations, for instance, appear to occur with greater frequency than do visual hallucinations.

A more general consequence of glottogenesis has to do with the institutionalization of social relationships, which Hinde (1987) represents as more or less equivalent to the evolution of human culture (a similar view is implicit in many if not most social anthropological definitions of culture: see Kroeber & Kluckhohn 1952 for a definitional review). However convincing may be the evidence from recent field work and laboratory experiments (related in previous chapters) that non-human primate behaviour is governed in complex ways by social cognition, those transactions between social partners which Humphrey (1976) has described as involving 'a two-way argument where each "player" must be ready to change his tactics – and maybe his goals – as the game proceeds . . . [where] the social gamesman, like the chess player, must be capable of a special sort of forward planning', reach their epitome in the social interaction of human beings capable of conversing not just about goals and desires of the moment but about the rules and categories which relate to conversations and interactions which they and others in the community have had or may be about to have, and on which current transactions may heavily depend. To suggest that language acquisition in development involves the learning of words which a child may then apply to things, relations and transformative actions, stringing them together in sequences of an order recognizable, significant, and not jarring to the ears of listeners in the community, involves the same sort of oversimplification as does description of learning by AI workers in terms of 'chunking' information and arriving at 'condition–action' rules (Chapter 4). Since details of process have yet to be worked out for either of these black box models of behaviour, we must not worry about the fact that they *are* black box models. They are still very useful for delineating general system relationships.

A recent review of how children learn language is found in Garton & Pratt (1989) from whom much of the following is taken. Like many such accounts, it relies on Chomsky's suggestion that the child has a Language Acquisition Device (LAD). We are aware of the controversy which has surrounded Chomsky's ideas since first publication, and like most evolutionists we reject the essentialist notion of a 'deep' grammatical structure embedded in the neural system at some moment in evolution by a single or a few mutations of cosmic proportion. But while Chomsky's ideas may not reveal much about the process of language evolution in hominids, they are nevertheless very informative about the product of that evolution, and it is the product which is their – and, at this point, our – main concern. The LAD is a black box 'wired to receive language input data and to output or generate [grammatical] sentences . . . The child has innate knowledge of the universal principles that govern the structure of language. These principles are resident in the mind of the child and are triggered by linguistic input' (Garton & Pratt 1989:18).

Here we can see the difference between language learning by a child and language invention by early man. In the former case the brain has to deal with a flow of incoming sounds which are themselves already organized into syntactical and grammatical structures; the child's task is first to tease apart the flow of sounds and form associations between the sounds and the surrounding environment, second to detect the organizational structure of the flow, and third to incorporate the rules of grammar and syntax into her or his own speech. In the case of early humans we can assume that there was a well established series of sounds used in communication, in association with gestures, but there was neither syntax nor grammar. As with the evolution of representational painting or ideas of the supernatural, the initial steps towards syntax and grammar were entirely novel human inventions, akin to the earlier hominid inventions such as the control of fire; in this case they were not technological–ecological but social.

In discussing the child's acquisition of spoken language, we are discussing an aspect of social interaction, because such complex language is difficult to acquire and its acquisition takes place, and has always taken place, in a social context (see studies, e.g., Davis 1948, of socially isolated children in whom no language development took place). Psychologists discussing child language learning have universally stressed the role of the mother (or equivalent caretaker) in the earliest stages of the process, and it is here that we encounter 'motherese' – the simplified language used by mothers to their young children. Garton & Pratt have tabulated the major characteristics of motherese, and their Table (Table 11.2) is interesting because it can suggest ideas of the ways in which evolutionary early language may have differed from fully developed language.

Bruner (1977, 1978) has pointed out deficiencies in the LAD concept, and has argued that the LAD alone does not suffice to explain speech acquisition. He introduced the notion of a Language Acquisition Support System (LASS). He maintained that 'through interaction with the mother, who guided and supported the child's emerging language, the child learned to talk, and learned the language of the socio-cultural and historical group in which he was growing up . . . [Thus] LAD can only function with LASS, as the social interaction format enables the child to learn the language' (Garton and Pratt 1989:45). It is the combination

Table 11.2.   *Major characteristics of motherese*

Category	Characteristic
Paralinguistic	High pitch Exaggerated intonation
Grammatical	Shorter utterance length Fewer verbs and modifiers Fewer subordinate clauses Fewer embeddings More content words Fewer function words
Discourse	More imperatives More interrogatives Speech more fluent and intelligible More repetitions

From Garton & Pratt, 1989:27.

of acquisition device within the child and support system provided by the mother or equivalent caretaker(s) which makes language learning and speech possible.

This is not a book about child development, and we need not ask just how LASS and LAD are conjoined – the rather too neat apposition of acronyms should not lull us into forgetting the difficulty of spelling out connections between a developmental support system, the structure of which can be rather easily discerned at least in broad outline, with its black box complement. We are looking at children's linguistic development to seek for clues about language evolution. What we can learn from the above is the importance not only of an evolved capacity to learn spoken language in the child but the equal importance of the social context, normally the presence of the (speaking) mother. Goodall has shown how chimpanzee offspring learn to use termite sticks by observing their mothers termite fishing (Hamburg 1968). The mother–child relationship is the ideal one for the transmission of cultural information. Language is such information. It is information the function of which is to enable the communication of information. Language learning by children is the most complex process of information learning known in the animal kingdom. An extremely difficult but richly rewarding task is being learned, and we see a high degree of motivation on both sides: the degree of attention of the mother towards her child and of the child towards its mother is very high and sustained.

Here, perhaps, is the appropriate place to consider the recent work of Savage-Rumbaugh (1990) on language learning by a pygmy chimpanzee, or bonobo, named Kanzi. Kanzi reportedly understands several hundred human words and has also mastered the basics of syntactical sentence structure, i.e. the difference between the use of a word as subject and object, etc. She does not learn simply to string familiar words together to earn rewards, as Terrace (1984; Terrace *et al.* 1979) believed that Nim and other subjects of 'ape language' studies had done, but can handle novel constructions of known words effectively.

Kanzi receives language both in its spoken form and in the form of a visual-graphic symbol system (pictographs on a portable card). She responds primarily by pointing to the relevant pictographs. In other words her comprehension of both speech and visual words is good, but her ability to produce replies is limited to the visual words. She attempts to imitate human speech but her capacity to do so is severely limited.

Neither food rewards nor shaping have been used to train Kanzi, who is highly motivated to understand communications directed to her and to communicate with others.

Is Kanzi a freak? By no means. Since the work with Kanzi, Sue Savage-Rumbaugh has gone on to co-rear a second young bonobo, this time with a common chimpanzee. Both have had the same experiences, and *both* have acquired large vocabularies and comprehension of spoken English. The bonobo consistently showed these skills earlier and in greater elaboration than the common chimpanzee. However, the fact that the latter has achieved so much indicates that we are constantly underestimating our nearest relatives, and that given the right environment from an early age, language comprehension results.

Where do apes fail? The answer, to date, is very clear: they cannot learn to produce speech. In so far as the brain contains some sort of LAD, it is not a device for comprehension alone, for apes can do that; rather, it must be a device for converting auditory inputs into equivalent vocal outputs. It may indeed be that specially evolved neural hardware is involved in this conversion process. We have known for decades that damage to Broca's or Wernicke's area of the brain can cause loss of speech function (Gordon 1990). It would be remarkable if something as complex as spoken language did not have dedicated neural substructures. Such substructures might be entirely developed during ontogeny, but on balance at present the most parsimonious hypothesis would seem to be that an evolved propensity for speech production exists in humans and not in any other species,[3] despite the impressive evidence for continuity in the case of speech comprehension.

Evolution of the LAD

Our excursion into the field of child development was directed at discovering, through studies of language learning, how to think about the *outcome* of language evolution. The answer we found was the LAD, a piece of neural hardware produced from a genetic template or, depending on one's view of it, a conceptualization the cognitive and neural character of which are unknown and perhaps unascertainable, but the function of which, in either case, is to coordinate incoming linguistic data and fashion it into speech.

Accepting that the LAD is the outcome of selection pressures on our ancestors leads us back to the question of what precisely we suppose to have been the advantages of spoken language which favoured its evolution as a central feature of human culture. The features listed in Table 11.1 represent one answer to this question. The question may also spur consideration of words as signifiers of meaning. G.H. Mead (1934) emphasized that words, because of their uniquely equal perceptibility to both speaker and hearer(s), have many advantages over non-verbal means of communication. If I say 'tree', then not only do I obtain instant confirmation that I have said 'tree' and no other word by hearing myself say it, but I also have confidence that my hearer(s) will have heard me say 'tree' and no other word. Whereas if I point to a tree, *I* may be sure it is a tree I am pointing to, but my communicatee may get the wrong idea and think I am pointing to something else entirely. Words thus add precision to our efforts to convey information (or for that matter to deceive one another).

Let us use this feature as a point of departure. When I say 'tree' I know my hearer hears 'tree' but how can I be sure my hearer will mean what I mean by 'tree'? The answer is that I can only be sure if both she and I are members of one and the same language community. What has happened is that in our respective childhoods we have both received the same word in the same contexts and have come to associate the word with the same class of objects.

In the absence of a language community, my saying the word 'tree' would have no communicative significance whatsoever. 'Tree' would be a conjunction of sounds but not a word. What would it take to turn a sound into a word in the absence of an existing language community? At minimum a hearer must make a connection between the sound and the concept the speaker has in mind when making the sound in the first place. There has to be an agreement about what de Saussure (1959) called the *significant* (signifier or 'word-image') and the *signifié* (signified or 'concept'). This agreement is negotiated when one learns a foreign language from an informant. The informant has to associate the word with the thing it stands for in the presence of and with the attention of the learner. The two have to agree on the connection between the sound(s) and the object (or relationship, quality, etc.). Thenceforth, they can communicate about that object in its absence by means of the sound(s), which has become a word for the learner as well as the informant. Although such negotiation doubtless has played an important role in the rise of languages, we can only speculate about its function in the early stages of language evolution, and to suggest that language originated as a product of semantic negotiation would seem explanatory mostly by way of metaphor, as, e.g., Rousseau's suggestion that human society was contrived by social contract. Nevertheless, a word is a Saussurean sign and is the basic unit of construction for a language. Its existence tends to have been taken too much for granted in the past, but recently linguists have asked the question why the Saussurean sign is at the foundation of languages (Hurford 1989). It may be that an analysis of the properties of the Saussurean sign can shed light on the evolution of the word, the evolution of the LAD, and the evolution of language of the human kind.

First we need to be clear about what is meant by a 'Saussurean sign'. It was represented by de Saussure (1959) as in Figure 11.1.

This diagram represents a relation between a concept and the sound-image of a word. The two arrows indicate that the sign or word 'arbor' is available for both transmission and reception. Thus a person saying 'arbor' conjures up the concept shown, and the person with the concept shown knows that the word for that concept is 'arbor'. But the arrows mean more than this. Generalizing from the single case of 'arbor', they also mean that the transmittor of the sound-image expects that, if a sound–image is used by another transmittor, it will have the same relationship to its concept as it has in cases when he himself uses it.

Figure 11.1.   Bidirectional Saussurean sign. (After Hurford 1989.)

LIVERPOOL
JOHN MOORES UNIVERSITY
AVRIL ROBARTS LRC
TEL. 0151 231 4022

Likewise a receiver, receiving a particular sound–image, expects that if he transmits it, it will stand for the same concept in other receivers as it has for himself.

This quality of bidirectionality of Saussurean signs is not God-given, nor is it a logical necessity, but just happens to be a property of human language as it has evolved. Hurford (1989) stresses that this often neglected feature is a basic ingredient of the LAD. He argues that it is so because of its superior communicative property over alternative possible kinds of signs. To test the efficacy of bidirectionality he designed a computer simulation to see which of three kinds of sign-users, Imitators, Calculators, or Saussureans, would achieve the greatest reproductive success after a number of generations; Saussureans won.

Hurford thus sees the evolution of the LAD as a result of biological mutations for factors involved in successful communication in the human environment, and the ability to incorporate Saussurean signs is one such factor. In his own words, the argument runs as follows:

> (1) Individuals who are more successful communicators enjoy a selective advantage and are more likely to reproduce than individuals who are worse communicators.
> (2) An innate Saussurean strategy for acquiring a communication system is superior to other conceivable strategies, and its possessors tend to enjoy a reproductive advantage over others, thereby increasing the prevalence of innate Saussurean strategists in the next generation.
> (3) Thereafter, over an evolutionary timespan, the Saussurean strategy displaces all rivals, and ends up being *the* strategy by which communication systems are naturally acquired. (p.194)

In his view, the LAD is a nice example of an evolved, genetically programmed system of neural hardware which has as its primary function the learning of Saussurean signs, and syntactical and grammatical rules. Initially, Hurford supposes, selection favoured the ability to communicate by bidirectional Saussurean signs, and subsequently by the combination of such signs into syntactical and grammatical structures enabling more complex forms of communication, with corresponding advantages to successful language users in terms of reproductive success.

We have dwelt on the ideas of Hurford because his is the only thoroughly worked out hypothesis for language evolution we have discovered which links in a systematic way de Saussure's observations with those of Chomsky to provide an explanation of the evolutionary background for the LAD, that logical construct postulated by Chomsky which has received at least partial validation from studies of child language learning. Should Chomsky's ideas, and those of the 'nativists', be disproven at some future time (i.e., should it be shown that there is no neural program 'designed' specifically to facilitate language learning, but it is after all managed by ordinary learning processes), then Hurford's ideas will fall at the same time.

Very recently, Györi (1990) has made a further contribution to ideas concerning the origins of linguistic signs. He challenges the idea that such signs were ever arbitrary, stressing instead that when signs were initially used to name phenomena, they were chosen on the basis of *features shared with the thing denoted*. And because there is historical continuity of linguistic signs from their first use to the present time, they still now cannot be called 'arbitrary'; our current words are, according to Györi, the much modified descendants of words which originally had something in common with the phenomena they denoted. Györi's analysis is based on his concept of cognition as an analogical process, in which 'everything new is compared to already acquired structures'.

5. Language and society

What do studies of child language learning and of language evolution, such as those we have been discussing, tell us about the evolution of human society? Much depends on what we see as the relationship between language and society. We argue here that the evolution of spoken language and the elaboration of social institutions went hand in hand. We have noted that it is in the mother–child relationship that language learning primarily takes place. On the assumption that kinship was the most important aspect of social life for our ancestors (as it is for all non-human primate species and for humans in many cultures) it seems likely that distinguishing between close kin (mother, father, sister, brother), kin at further removes (grandfather, grandmother, sister's son, mother's brother's daughter), and non-kin (individuals of other lineage) may have been the primary context of language use during human evolution. We suggest this because information about social relationships can be conveyed with far greater precision through language than without it, and such information must count among the most important in a lineage-based primate species.

We have referred previously to Dasser's (1988) work, showing that macaques can identify mother–infant pairs in a series of well controlled tests using photographs of individuals they know well. In other words, the beginnings of recognition of kin relationships are to be seen in the non-verbal primates. This fact suggests that there are advantages to being able to recognize such relationships, and it is not hard to suggest reasons, for example it will help to avoid unnecessary involvement in agonistic encounters, and it will help to make speedy decisions concerning appropriate assistance to kin or other associates when conflict is unavoidable. Such decisions are by no means limited to the here-and-now for non-human primates. De Waal (1989c) has shown how prolonged is the recollection of social insult for chimpanzees, who exact 'revenge' long after the presumed causative event. In this temporal extension of social action governed by a single retaliatory decision, with iterative applications, we see the bedrock pattern of the human institution of the *feud*, to which we have also previously referred, which requires a degree of mobilization of kin support inconceivable in a primate species lacking fully developed language, for human beings language has greatly extended the temporal arena of decision-making as well as, in every dimension, the categories to which decisions are applied.

6. Naming

Naming (see also Chapter 10) and feuding are inseparable: it is the name of the rival lineages which distinguish them, and it is the name of each individual member of a feuding lineage which embroils him or her in social action, acting as an inescapable obligation, or singling him or her out as a legitimate target. By naming we here refer to lineage names, or clan names, or family names, and such names are found throughout the world's cultures. Their use is to provide social markers, recognizable divisions of society, through which marriage rules can operate, economic alliances of reciprocal kinds can be constructed (Layton 1989), and disputes (including feuds) can be channelled.

The word 'name' can also be used in a general sense for the naming of anything. Everything of which it is possible to talk has a name. A large part of language consists of names, for objects, ideas and relationships. These are the words, or Saussurean signs, referred to above. In the field of kinship, we often refer to close kin by the name of the relationship we have with them rather than by their proper names, thus we call out 'Mum!', or 'Dad!' There are names for relationships which are exclusively for intra-lineage use, there are individual names such as Tom or Mary for intra- and inter-lineage use, and there are names primarily for inter-lineage use, as for instance when one talks of the Jones's or the members of the Emu clan. Naming of these kinds must be very ancient; names are present in the earliest texts – the epic of Gilgamesh, the Dead Sea Scrolls, the Old Testament, ancient Egyptian papyri, Homer. Naming has gone on unabated ever since. In the West Indies at the present time and perhaps elsewhere, children are given proper names according to some prominent event which happened at the time of their birth, e.g., Hurricane or Christmas. In Uganda one of us encountered a silent uncommunicative man whose name was 'Hapanawatu' meaning 'no people'. Elsewhere names may be arbitrary (e.g., Richard), or they may be linked to a trade past or present (e.g., Glover), or they may be linked to property ownership past or present (e.g., the Earl of Cumberland), or they may, as in Iceland, express whose child the named person is. However varied names may be, all people have names, naming is universal, and in many countries is subject to particular rules and ceremonies. In the Christian world we have the ritual of baptism, by which the infant is not just made a member of the Christian community but is given a name or names. In many traditional societies a child is not named until some time after birth and has no social existence until then. Naming is a very elementary part of human life.

Lineage names

If the naming of an individual is a primary part of the socialization process, giving a recognizable and communicable identity to the new member of society, then the family name, lineage name, or clan name is equally or even more important, for it gives symbolic expression to the entity recognized in the culture in question as a primary kinship unit. If rhesus or Japanese monkeys could speak they would have lineage names, for matrilines are an important part of their social life. In studying such primates, students often give them individual names, not infrequently using a common prefix or suffix for matriline membership (e.g., Glance 6476, a female Japanese monkey, born in 1976 to the Glance Matriline of Arashiyama East troop, who in her youth initiated 'stone play' at Arashiyama (Huffman 1984)).

What is such a lineage name for? It represents a class or group or unit of individuals, distinct from other such classes or units. Totemic lineages, as found in Aboriginal Australia or in Northwest Coast American Indians, have animals for the names of their lineages or clans. Having named the lineage or clan, rules of interaction between clans can be formulated. Bears do not marry Bears, they marry Crows. Or: Bears and Crows are closer to each other than they are to Sockeye Salmon because their ancestor Snake created them, whereas Sockeye Salmon were created by Dog. So Bears and Crows should marry Sockeye Salmon rather than each other, but if need be they may marry each other, but Bears must never marry Bears.

The details are of no significance here. What is important is that naming can be a form of social 'chunking'. Instead of having to work out just how another individual may be related to oneself, or not, we can be guided by the name. The name tells us whether he or she is a relative, a potential marriage partner, or a rival. We are back to our AI parallel, and we can see that just as primates have always been good at chunking in the field of social relations, taking on board a lot of social information and using it to make decisions about how to conduct their social affairs, so the evolution of language by humans has enabled them to extend their field of social expertise into the furthest reaches of kinship and, with the invention of marriage, affinity. Language use enabled the refinement of social cognition, widening the field of allies and coalition partners available in time of need, and it is for this reason that those individuals and families in which efforts were made to name relatives gained a selective advantage over those who did not or could not.

This advantage of extending kinship links is a direct development of the situation in primates in which larger lineages come to dominate smaller ones (Samuels, *et al.* 1987). That study of baboon lineages showed how one lineage, because of a depletion of its members, lost its dominant status (Chapter 9). Human lineages are often also in competition with one another for property, movable wealth, and other scarce resources. A lineage under pressure from a rival lineage calls on other relatives from further afield to defend itself. Tribes such as the LoWiili (Goody 1956) or the Nuer (Evans-Pritchard 1940), in which there is no central authority but power is distributed through a segmentary lineage system, succeed in managing their political affairs by an extensive system of checks and balances between lineages and clans. In the competition between early human hunting–gathering bands, political alliances with kin in other bands may have been the principle by which some lineages were able to secure long-term security of food supply.

Alternative models

We have emphasized the naming of lineages in the creation of the institutions of society. From the perspective of primatology, the three- or four-generational lineage (in which individuals living out a 'normal' lifespan are likely to interact) is arguably the unit of most functional significance beyond the level of the individual. Studies of human societies by social anthropologists have, however, emphasized additional kinds of classes and divisions. There are age-sets in East African pastoralists, there is the distinction between matrilineal and patrilineal kin, there is blood-brotherhood by which a form of kinship is created, there is fictional kinship, and there is the division between cognatic (genetic) and affinal kin (through marriage). Could the re-shaping of human society by means of language have been based on principles other than the principle of lineal relationship?

A very different approach to the question of the construction of a human kind of society is taken by Allen (1982, 1986, 1989), a social anthropologist who does not start from a Darwinian perspective or use data from primatology as a baseline for his analysis. The most notable thing about Allen's scheme for the evolution of society is that it does not start with pre-existent lineage structure, around which further developments occur, but avoids lineages altogether, regarding them as an unnecessary complication. Allen calls his theory of the origin of human society 'tetradic' because his model has four *sections* (See Figure 11.2).

The terminology of this kind of society (which resembles that of some societies of Australian aborigines) is unremarkable. What is perhaps remarkable about it is that Allen shows that with a simple marriage rule, namely that a person must only marry a cross cousin, this social structure can continue into perpetuity as a perfectly functioning whole, avoiding incest. In other words, he has shown that it is not logically necessary to start one's model of early human society with lineages, and to see the institution of cross-cousin marriage as a link-up of lineages. This form of marriage requires no more than a two-generational structure, alternate generations being equated with each other, i.e., the third generation is, functionally and in other ways, the exact equivalent of the first, the second with the fourth, etc., from the point of view of the model.

We should note, therefore, that alternative theoretical possibilities to a lineage-based society exist, are highly logical in their construction and demand a minimum of social engineering to work. According to Allen (1989), lineage-based societies are a subsequent development of section-based, tetradic societies. He writes, 'Humanity started off with terminologies which can be envisaged as the result of superimposing three types of equation. These are 1. classificatory equations, 2. prescriptive

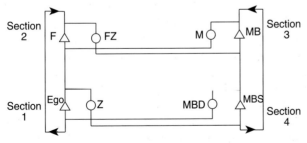

Figure 11.2.   Tetradic society and terminology. After Allen (1986).

equations, and 3. alternate generation equations. In the course of prehistory and history the dominant trend has been towards the rupture of these equations' (1989:178).

We note Allen's scheme, based on elemental logic, with interest because it represents a challenge to our ideas, being based on the working out of a simple though subtle scheme presumed to be the result of the cogitations of our earliest, society-constructing ancestors. We have not adopted this approach, because from a Darwinian perspective we have preferred to argue in terms of continuity, and have thus allowed the forms of social structure and the kinds of relationships found in non-human primates to guide our ideas. The result has been that in our model lineages have been stressed from the outset, as has lineage recruitment and lineage competition, three items missing from Allen's model. Indeed, the tetradic model has no competition of any kind. Despite this, the two schemes do have one thing in common, namely that they consist of entities or classes (lineages, sections) which require designating in some unambiguous way. Allen suggests (1986) that this could be done by different kinds of body painting or (1989) hairstyles. So too could lineages. That would be to go back to a pre-linguistic stage. At some time, such simple kinds of distinctions failed to cope, and we have argued that social diversification provided the context for the evolution of language.

6. Social rules

With the evolution of language, kinship became a game. Not a game in the sense that it was a leisure pursuit, but a game in the sense of Wittgenstein's language games or in the sense of game theory, in other words an ordering and reordering of relationships, over time, to produce novel arrangements and meanings. The primary feature of any game is its rules. Rules have been emphasized by many anthropologists as the central feature of human societies, and Fortes (1983), in his last published work, made rules the crucial element in his description of the emergence of human society.

The late Meyer Fortes was one of the few social anthropologists to have seriously considered the question that forms the subject of this book, namely how the institutions of human society might have arisen out of pre-human primate society. Rather more than we have, he emphasized language as man's unique criterion. Culture, he wrote, constrains humans through language, by 'the sovereignty of rules in social life' (p.4). He distinguished between prescriptive and proscriptive rules, especially in the field of human reproduction, and between the etic and emic aspects of rules – the rules as determined by outsiders such as anthropologists, and as understood by insiders, the rule-followers themselves. Rules, he argued, have juridical and moral force, are backed by authority, and are accepted and changed by consensus.

'Fatherhood', according to Fortes, was 'a creation of society', and was 'man's central rule-generating institution' (p.20). By fatherhood he meant either the *pater* of patrilineal societies, the *pater* + *avunculus* of matrilineal societies, or a corporate (male) group. Fortes was aware of the descriptions of kinship in non-primate societies, but was unwilling to consider that non-human primates could have kinship in the same sense of the word that humans have kinship. For Fortes, human kinship is based on what he calls 'prescriptive altruism' (p.24), which is not to be equated with the altruism found in animal kinship systems because it is based on rules and not on shared genetic constitution or behavioural alliances. He goes to some lengths to contrast biological and prescriptive altruism. Rules, according to Fortes, exist to maintain mutual trust, and they are not concerned with biological fitness, whereas biological altruism is, ultimately, concerned with reproductive fitness.

This paper by Fortes constitutes a remarkable synthesis of ideas from ethology, sociobiology, and social anthropology, and in many ways represents what is still the position of many anthropologists. We hope that our book has advanced beyond that position and left it untenable. First, our argument from continuity has shown that it is unreasonable to posit a sudden break or leap from animal kinship to human kinship at the time of language evolution and the emergence of rule-based society. Language itself did not evolve suddenly but probably moved slowly from the stage of Saussurean signs (gestural, to begin with, if we adopt Hewes's two-stage theory) to more complex syntactical and grammatical stages. The infiltration of words and then sentences into the field of kinship must have been halting and slow to begin with, quickening rapidly after (or, more likely, in close conjunction with) the vocalization of language.

Second, rules as jural prescriptions and proscriptions demand fully evolved language and must therefore be seen as a development which would have followed on from language evolution, so that language would have been incorporated into kin relationships before rules were devised to govern those relationships.

Third, Fortes' argument that rules were an outcome of the creation of fatherhood by society seems perhaps his weakest one. His contrast between maternal love and paternal authority, based undoubtedly on his lifelong studies of modern human societies, seems naive. Why should rules have been devised by one sex in particular? It seems to us that the creation of rules would occur as and when the need arose, by an initial consensus of two or three persons of either or both sexes, and later spreading to others, again by consensus, according to perceptions of the advantages to be gained from those rules.

Fortes' claim that prescriptive altruism, i.e., altruism based on linguistic rules, needs to be sharply distinguished from biological altruism, i.e., altruism based on considerations of Darwinian fitness, falls straight into the error of confusion of levels of analysis. Prescriptive altruism is the kind of cooperative, friendly, helpful behaviour which arises from some such rule as 'Do as you would be done by'. Such a rule exists in many societies as a guide to action. It exists in the here and now, as a rule one can learn and can either abide by or ignore. On the other hand, the fitness differentials which have led to the evolution of biological altruism, whether kin-based or reciprocal, are not rules of the here and now; they are ultimate not proximate determinants of action. For this reason the argument that one precludes or is incompatible with the other does not stand. Both prescriptive rules for altruistic actions and biological causes of altruistic behaviour can co-exist as determinants. Alexander (1987) in particular has argued that the rules of consensus, altruism, and morality found in modern human societies are based on biological altruism.

These criticisms should not be taken simply as an attempt to refute Fortes' position because we do not agree with Fortes. They are intended rather to emphasize the continuity of non-human and human social life. The emphasis we have placed on cognition in primate social life has led on, via a focus on kinship, to the view of how a slow but steady transformation of society could come about with the evolution of language. All the developments we have described can be encompassed within a Darwinian framework. Language evolution has brought about modifications of the hardware of human brains. Those brains, using language, have brought about modifications of human societies. Underlying such modifications has been the competition between individuals and between lineages. Individuals have pursued simultaneously their own and collective lineal interests. Social life in humans remains, in part competitive. It is this element of competition which is lacking in so much of anthropological thinking and description, perhaps because of the influence of Durkheim's emphasis on the 'social contract'. As this book has shown, we do not believe in 'nature red in tooth and claw' but neither do we accept that the principle of social organization is harmony. Both positions are simplistic in the extreme. The truth is that competitive and cooperative strands are interwoven in primate action at all levels, from that of the individual young monkey interacting with siblings and playmates in his or her natal group to that of the human child inheriting citizenship in a nation-state. The study of human society begins with the analysis of how the institutionalized competition and cooperation characteristic of humankind has grown out of the already rich competitive and cooperative relationships and structures seen in non-human primates.

Notes

1. Our discussion in this section is drawn from Milo & Quiatt (in press a,b). For a fuller review of the evidence from archaeology and comparative anatomy bearing on language evolution see those publications.
2. Wynn's rigorous analysis of spatial competence, 1989a, is focused exclusively on manufacturing process as reflected in flakes removed from worked stone. This narrow focus allows him to calculate *minimum* cognitive requirements for material manufacture. That manufacture and use of, say, Acheulean handaxes, may not occur at all except in a complex society of highly intelligent language users is a distinct possibility, but Wynn properly avoids speculation on these matters.
3. The case of birds that learn words and 'speak' needs a mention. Such 'parroting' has been shown to be characteristic of species which learn their calls in the wild. Some individual parrots have learned to speak a hundred or more words. The process is, however, limited; there is no evidence of understanding of reference or agency, no comprehension of self, and no comprehension of use of syntax.

References

Alexander, R. D. (1979). *Darwinism and Human Affairs*. Seattle: University of Washington Press.

Alexander, R. D. (1987). *The Biology of Moral Systems*. New York: Aldine de Gruyter.

Allen, N. J. (1982). A dance of relatives. *Journal of the Anthropological Society*, **13**, 139–46.

Allen, N. J. (1986). Tetradic theory: An approach to kinship. *Journal of the Anthropological Society*, **17**, 87–109.

Allen, N. J. (1989). The evolution of kinship terminologies. *Lingua*, **77**, 173–85.

Altmann, J. (1974). Observational study of behaviour: sampling methods. *Behaviour*, **49**, 227–67.

Altmann, S. A. (1965). Sociobiology of rhesus monkeys. II. Stochastics of social communication. *Journal of Theoretical Biology*, **8**, 490–522.

Baker, P. T. & Little, M. A. (eds.) (1976). *Man in the Andes: Multidisciplinary Study of High Altitude in Quechua*. Stroudsburg, Pennsylvania: Dowden, Hutchinson & Ross.

Barash, D. (1977). *Sociobiology and Behavior*. New York: Elsevier.

Barlow, G. W. (1988). Monogamy in relation to resources. In *The ecology of Social Behavior*, ed. C. N. Slobodchikoff, pp. 55–79. New York: Academic Press.

Beach, F. A. (1950). The Snark was a boojum. *American Psychologist*, **5**, 115–24. (Presidential address delivered to the Division of Experimental Psychology of the American Psychological Association., September 7, 1949)

Bekoff, M. (1981). Mammalian sibling interactions: genes, facilitative environments, and the coefficient of familiarity. In: *Parental Care in Mammals*, ed. D. J. Gubernick & P. H. Klopfer, pp. 307–46. New York: Plenum.

Bernstein, I. S. & Ehardt, C. L. (1986). The influence of kinship and socialization on aggressive behaviour in rhesus monkeys (*Macaca mulatta*). *Animal Behaviour*, **34**, 739–47.

Bernstein, I. S. & Sharpe, L. G. (1966). Social roles in a rhesus monkey group. *Behaviour*, **26**, 91–104.

Bickerton, D. (1990). *Language and Species*. Chicago: Chicago University Press.

Binford, L. E. (1981). *Bones: Ancient Men and Modern Myths*. New York: Academic Press.

Boden, M. A. (1987). *Artificial Intelligence & Natural Man*. New York: Basic Books.

Boelkins, R.C. & Wilson, A.P. (1972). Intergroup social dynamics of the Cayo Santiago rhesus (*Macaca mulatta*) with special reference to changes in group membership by males. *Primates*, **13**, 125–40.

Boesch, C. & Boesch, H. (1986). Film presentation to the International Society of Human Ethologists.

Boesch, C. & Boesch, H. (1990). Tool use and tool making in wild chimpanzees. *Folia Primatologica*, **54**, 86–99.

Borgerhoff-Mulder, M. (1987a). On cultural and reproductive success: Kipsigis evidence. *American Anthropologist*, **87**, 617–34.

Borgerhoff-Mulder, M. (1987b). Adaptation and evolutionary approaches to anthropology. *Man*, **22**, 25–41.

Bott, E. (1957). *Family and Social Network*. London: Tavistock.

Boucher D. H. (1977). On wasting parental investment. *American Naturalist*, **111**, 786–88.

Boyd, R. & Richerson, P. J. (1985). *Culture and the Evolutionary Process*. Chicago: University of Chicago Press.

Brain, C. K. (1981). *The Hunters Or the Hunted? An Introduction to African Cave Taphonomy*. Chicago: Aldine.

Bramblett, C. A. (1976). *Patterns of Primate Behavior*. Palo Alto, California: Mayfield.

Brockelman, W. Y. (1984). Social behaviour of gibbons: introduction. In: *The Lesser Apes: Evolutionary and Behavioural Biology*, ed. H. Preuschoft, D. J. Chivers, W. Y. Brockelman, & N. Creel, pp. 285–90. Edinburgh: Edinburgh University Press.

Brockelman, W. Y. & Treesucon, U. (1986). Observations on social changes in the gibbon *Hylobates lar* and *H. lar – H. pileatus* hybrids in Thailand. *Primate Report*, **14**:72. (Abstract only).

Bruner, J. S. (1977). Early social interaction and language development. In *Studies in Mother–Child Interaction*, ed. H. R. Schaffer. London: Academic Press.

Bruner, J. S. (1978). Learning how to do things with words. In *Human Growth and Development*, ed. J. Bruner & A. Garton. Oxford: Clarendon Press.

Bunn, H. T. & Kroll, E.M. (1986). Systematic butchery by Plio/Pleistocene hominids at Olduvai Gorge, Tanzania. *Current Anthropology*, **27**, 431–52.

Byrne, R. & Whiten, A. (1985). Tactical deception of familiar individuals in baboons (*Papio ursinus*). *Animal Behaviour*, **33**, 669–673.

Byrne, R. & Whiten, A. (eds). (1988). *Machiavellian Intelligence: Social Expertise and the Evolution of Intellect in Monkeys, Apes, and Humans*. Oxford: Clarendon Press.

Byers, J. A. & Bekoff, M. (1986). What does 'kin recognition' mean? *Ethology*, **72**, 342–5.

Byers, J. A. & Bekoff, M. (1991). Development, the conveniently forgotten variable in 'true kin recognition'. *Animal Behaviour*, **41**, 1088–90.

Candland, D. K. & Leshner, A. I. (1974). A model of agonistic behavior: Endocrine and autonomic correlates. In *Limbic and Autonomic Nervous System Research*, ed. L.V. DiCara, pp. 137–163. New York: Plenum.

Carpenter, C. R. (1940). A field study in Siam of the behavior and social relations of the gibbon. *Comparative Psychology Monographs*, Vol. 16, No. 5 (December).

Carpenter, C. R. (1942). Societies of monkeys and apes. *Biological Symposia*, **8**, 177–204.

Carrithers, M. (1985). *The Category of the Person*. Cambridge: Cambridge University Press.

Carrithers, M. (1990). Why humans have cultures. *Man*, **25**, 189–206.

Cartmill, M. (1974). Rethinking primate origins. *Science*, **184**, 436–43.

Chadwick-Jones, J. K. (1987a). Social psychology and primatology: proximate explanations. *Ethology*, **74**, 164–9.

Chadwick-Jones, J. K. (1987b). A social psychology of non-human primates: a developing field. *New Ideas in Psychology*, **5**(1), 111–15.

Chadwick-Jones, J. K. (1989a). Presenting and mounting in non-human primates: theoretical developments. *Journal of Social Biological Structure*, **12**, 319–33.

Chadwick-Jones, J. K. (1989b) Baboon charades. *The Psychologist*, **2**, 58–61.

Chagnon, N. A. (1979). Is reproductive success equal in egalitarian societies? In *Evolutionary Biology & Human Social Behavior: An Anthropological Perspective*, ed. N. A. Chagnon & W. Irons, pp. 374–401. N. Scituate, MA.: Duxbury Press.

Chagnon, N. A. (1988). Male Yanomamo manipulations of kinship. In *Human Reproductive Behaviour*, ed. L. Betqig, *et al*. Cambridge: Cambridge University Press.

Chalmers, N. (1979). *Social Behaviour In Primates*. London: Edward Arnold.

Chance, M. R. A. (1962). Social behaviour and primate evolution. In *Culture and the Evolution of Man*, ed. M. F. Ashley Montagu. New York: Oxford University Press.

Chance, M. R. A. (1967). Attention structure as the basis of primate rank orders. *Man*, **2**, 503–18.

Chance, M. R. A. (ed.). (1988). *Social Fabrics of the Mind*. London: Lawrence Erlbaum.

Chapais, B. (1983). Dominance relatedness, and the structure of femal relationships in rhesus monkeys. In *Primate social relationships: an integrated approach*, ed. R. A. Hinde, pp. 209–17. Oxford: Blackwell.

Chatwin, B. (1987). *The Songlines*. New York: Viking.

Cheney, D. L. (1984). Category formation in vervet monkeys. In: *The Meaning of Primate Signals*, ed. R. Harré & V. Reynolds, pp. 58–72. Cambridge: Cambridge University Press.

Cheney, D. L. & Seyfarth. R. M. (1980). Vocal recognition in free-ranging vervet monkeys. *Animal Behaviour*, **28**, 362–67.

Cheney, D. L. & Seyfarth, R. M. (1982a). How vervet monkeys perceive their grunts: field playback experiments. *Animal Behaviour*, **30**, 739–51.

Cheney, D. L. & Seyfarth, R. M. (1982b). Recognition of individuals within and between groups of free-ranging vervet monkeys. *American Zoologist*, **22**, 519–29.

Cheney, D. L. & Seyfarth, R. M. (1985). Vervet monkey alarm calls: manipulation through shared information? *Behaviour*, **93**, 150–66.

Cheney, D. L. & Seyfarth, R. M. (1986). The recognition of social alliances by vervet monkeys. *Animal Behaviour*, **34**, 1722–31.

Cheney, D. L. & Seyfarth, R. M. (1988). Assessment of meaning and the detection of unreliable signals by vervet monkeys. *Animal Behaviour*, **36**, 477–86.

Cheney, D. L. & Seyfarth, R. M. (1990). *How Monkeys See the World. Inside the Mind of Another Species*. Chicago: University of Chicago Press.

Cheney, D. L., Seyfarth, R. M. & Smuts, B. (1986). Social relationships and social cognition in nonhuman primates. *Science*, **234**, 1361–6.

Chism, J. B. & Rowell, T. E. (1986). Mating and residence patterns of male patas monkeys. *Ethology*, **72**, 31–9.

Chivers, D. J. & Raemaekers, J. J. (1980). Long-term changes in behaviour. In: *Malayan Forest Primates*, ed. D. J. Chivers, pp. 209–60. New York: Plenum Press.

Chomsky, N. (1962). Explanatory models in linguistics. In *Logic Methodology and the Philosophy of Science*, ed. E. Nagel, P. Suppes & A. Tarsky. Stanford: Stanford University Press.

Chomsky, N. (1967). The general properties of language. In *Brain Mechanisms Underlying Speech and Language*, ed. F. L. Darley pp. 73–88. New York: Grune & Stratton.

Chomsky, N. (1970). Problems of explanation in linguistics. In *Explanation in the Behavioral Sciences*, ed. R. Borger & F. Cioffi pp. 425–51. Cambridge: Cambridge University Press.

Clark, W. E. Le Gros (1959). *The Antecedents of Man*. Edinburgh: Edinburgh University Press.

Clark, W. E. Le Gros (1970). *History of the Primates*. London: British Museum of Natural History.

Clarke, W. E. Le Gros (1971). *The Antecedents of Man*, 3rd edn. Edinburgh: Edinburgh University Press.

Clifford, J. & Marcus, G. E. (eds.). (1986). *Writing Culture: The Poetics and Politics of Ethnography*. Berkeley: University of California Press.

Clutton-Brock, T. H. (1989). Female transfer and inbreeding avoidance in social mammals. *Nature*, **337**, 70–2.

Clutton-Brock, T. H. & Albon, S. D. (1982). Parental investment in male and female offspring in mammals. In *Current Problems in Sociobiology*, ed. King's College Sociobiology Group, pp. 223–48. Cambridge: Cambridge University Press.

Clutton-Brock, T. H. & Harvey, P. H. (1976). Evolutionary rules and primate societies. In *Growing Points In Ethology*, ed. P. P. G. Bateson & R. A. Hinde, pp.195–237. Cambridge: Cambridge University Press.

Clutton–Brock, T. H., Albon, S. D., Gibson, R. M. & Guinness, F. E. (1979). The logical stag: adaptive aspects of fighting in red deer (*Cervus nelaphus* L.). *Animal Behaviour*, **27**, 211–25.

Colmenares, F. & Rivero, H. (1986). A conceptual model for analysing social interactions in baboons: a preliminary report. In *Quantitative Models In Ethology*, eds. P. W. Colgan & R. Zayan, pp. 63–80. Toulouse: Privat, I.E.C.

Colvin, J. (1986). Proximate causes of male emigration at puberty in rhesus monkeys. In *The Cayo Santiago Macaques*, ed. R. G .Rawlins and M. J. Kessler. Albany, NY: SUNY Press.

Conference on Group Processes, (1954–58). *Group Processes Transactions*. Josiah Mary Foundation. Vols. 1–5.

Conroy, G. C. (1990). *Primate Evolution*. New York: W. W. Norton.

Crook, J. H. (1964). The evolution of social organization and visual communication in the weaver birds (*Ploceinae*). *Behaviour Monograph 10*. Leiden: Brill.

Crook, J. H. (1965). The adaptive significance of avian social organizations. *Symposia of the Zoological Society of London*, **14**, 181–218.

Crook, J. H. (1970). Social organization and the environment: aspects of contemporary social ethology. *Animal Behaviour*, **18**, 197–209.

Crook, J. H. (1989). Introduction: socioecological paradigms, evolution and history; perspectives for the 1990s. In *Comparative Socioecology: The Behavioural Ecology of Humans and Other Mammals*, ed. V. Standen & R.A. Foley pp. 1–36. Oxford: Blackwell Scientific Publications.

Crook, J. H. & Gartlan, J. S. (1966). Evolution of primate societies. *Nature*, **210**, 1200–3.

Dacey, F. (1973). Some questions about spatial distributions. In *Directions In Geography*, ed. R. V. Chorley. London: Methuen.

Dasser, V. (1988). A social concept in Java monkeys. *Animal Behaviour*, **36**, 225–30.

Datta, S. B. (1986). The role of alliances in the acquisition of rank. In *Primate Ontogeny, Cognition and Social Behaviour*, ed. J. G. Else & P. C. Lee pp. 219–25. Cambridge: Cambridge University Press.

Datta, S. B. (1989). Demographic influences on dominance structure among female primates. In *Comparative Socioecology: The Behavioural Ecology of Humans and Other Mammals*, ed. V. Standen & R. A. Foley, pp. 265–84. Oxford: Blackwell Scientific Publications.

Davenport, R. K., Jr. & Rogers, C. M. (1970). Intermodal equivalence of stimuli in apes. *Science*, **168**, 279–80.

Davis, K. (1948). *Human Society*. New York: Macmillan.

Dawkins, M. S. & Guilford, T. (1991). The corruption of honest signalling. *Animal Behaviour*, **41**, 865–73.

Dawkins, R. (1976). *The Selfish Gene*. Oxford: Oxford University Press.

Dawkins, R. (1979). Twelve misunderstandings of kin selection. *Zeitschrift für Tierpsychologie*, **51**, 184–200.

Dawkins, R. (1980). Good strategy or evolutionarily stable strategy? In *Sociobiology: Beyond Nature/Nurture?* ed. G. W. Barlow & J. Silverberg, pp. 331–67. Boulder: Westview Press.

Dawkins, R. (1982). Replicators and vehicles. In *Current Problems In Sociobiology*, ed. King's College Sociobiology Group, pp. 45–64. Cambridge: Cambridge University Press.

Dawkins, R. (1986). *The Blind Watchmaker: Why the Evidence of Evolution Reveals a Universe Without Design*. New York/London: W. W. Norton.

Dawkins, R. (1989). Taking the high ground. Review of 'Arguments on evolution: a paleontologist's perspective', by Antoni Hoffman. *Nature*, **340**, 25.

Dawkins, R. & Krebs, J. R. (1978). Animal signals: information or manipulation. In *Behavioural Ecology: An Evolutionary Approach*, ed. J. R. Krebs & N. B. Davies, pp. 282–309. Oxford: Blackwell Scientific Publications.

Deag, J. M. and Crook, J. H. (1971). Social behaviour and 'agonistic buffering' in the wild Barbary macaque. *Macaca sylvana* L. *Folia Primatologica*, **15**, 183–200.

Dennett, D. C. (1983). Intentional systems in cognitive ethology: the 'Panglossian paradigm' defended. *Behavioral and Brain Sciences*, **6**, 343–90 (W/BBS commentary).

Dittus, W. P. J. (1980). The social regulation of primate populations: a synthesis. In *The Macaques*, ed. D. G. Lindburg, pp.263–86. New York: Van Nostrand.

Dittus, W.P.J. (1986). Sex differences in fitness following a group takeover among Toque Macaques: testing models of social evolution. *Behavioural Ecology and Sociobiology*, **19**, 257–66

Dreyfus, H. L. & Dreyfus, S. E. (1988). Making a mind versus modeling the brain: artificial intelligence back at a branchpoint. In *The Artificial Intelligence Debate: False Starts and Real Foundations*, ed. S. R. Gravbard, pp. 15–43. Cambridge, Massachusetts: MIT Press.

Drickamer, L.C. & Vessey, S.H. (1973). Group changing in free-ranging male rhesus monkeys. *Primates*, **14**, 359–68.

Drucker, P. (1951). *The Northern and Central Nootkan Tribes. Bulletin 144.* Washington, DC: The Smithsonian Institution.

Dumont, L. (1970). Homo Hierarchicus: *The Caste System and Its Implications.* London: Weidenfeld and Nicholson.

Dunbar, R. I. M. (1984). *Reproductive Decisions: An Economic Analysis of Gelada Baboon Social Strategies.* Princeton, New Jersey: Princeton University Press.

Dunbar, R. I. M. (1988). *Primate Social Systems.* Ithaca, New York: Cornell University Press.

Dunbar, R. I. M. (1989). Social systems as optimal strategy sets: the costs and benefits of sociality. In *Comparative Socioecology: The Behavioural Ecology of Humans and Other Mammals*, ed. V. Standen & R. A. Foley pp. 131–49. Oxford: Blackwell Scientific Publications.

Easley, S. P., Coelho, A. M., Jr. & Taylor, L. L. (1989). Allogrooming, partner choice, and dominance in male anubis baboons. *American Journal of Physical Anthropology*, **80**, 353–68.

Eberhard, W. G. (1985). *Sexual Selection and Animal Genitalia.* Cambridge, Massachusetts: Harvard University Press.

Eisenman, L. M. (1978). Vocal communication in primates. In *Sensory Systems of Primates*, ed. C. R. Noback, pp. 93–108. New York: Plenum Press.

Eldredge, N. (1985a). *Time Frames. The Rethinking of Darwinian Evolution and the Theory of Punctuated Equilibria.* New York: Simon & Schuster.

Eldredge, N. (1985b). Evolutionary tempos and modes: a paleontological perspective. In *What Darwin Began: Modern Darwinian and Non-Darwinian Perspectives on Evolution*, ed. L. R. Godfrey, pp. 113–37. Boston: Allyn and Bacon.

Ellefson, J. O. (1968). Territorial behavior in the common white-handed gibbon, *Hylobates lar* Linn. In *Primates: Studies in Adaptation and Variability*, ed. P.C. Jay, pp. 180–99. New York: Holt, Rinehart & Winston.

Emlen, S. T., & Oring, L. W. (1977). Ecology, sexual selection, and the evolution of mating systems. *Science*, **197**, 215–23.

Epple, G. (1986). Communication by chemical signals. In *Comparative Primate Biology, Vol. 2A: Behavior, Conservation, and Ecology*, ed. G. Mitchell & J. Erwin, pp. 531–580. New York: Alan R. Liss

Epple, G. & Moulton, D. (1978). Structural organization and communicatory functions of olfaction in nonhuman primates. In *Sensory Systems of Primates*, ed. C. R. Noback, pp. 1–22. New York: Plenum Press.

Essock-Vitale, S. & Seyfarth, R. M. (1987). Intelligence and social cognition. In *Primate Societies*, ed. B. Smuts *et al.*, pp. 452–61. Chicago: University of Chicago Press.

Ettlinger, G. (1970). Cross-modal transfer of training in monkeys. *Behaviour*, **16**, 56–65.

Evans-Pritchard, E. E. (1940). *The Nuer.* Oxford: Clarendon Press.

Falk, D. (1984). The petrified brain. *Natural History*, **93**(9), 36,38–39.

Falk, D. (1987). Hominid paleoneurology. *Annual Review of Anthropology*, **16**, 13–20.

Fedigan, L. M. (1982). *Primate Paradigms: Sex Roles and Social Bonds.* Montreal: Eden Press.

Ferster, C. B. (1964). Arithmetic behavior in chimpanzees. *Scientific American*, **210**, 98–106.

Ferster, C. B. & Hammer, C. E., Jr. (1966). Synthesizing the components of arithmetic behavior. In *Operant Behavior: Areas of Research and Application*, ed. W. K. Honig, pp. 634–76. New York: Appleton-Century-Crofts.

Fleagle, J. G. (1988). *Primate Adaptation and Evolution.* New York: Academic Press.

Fletcher, D. J. C. & Michener, C. D. (ed.). (1987). *Kin Recognition in Animals.* New York: Wiley Interscience.

Fobes, J. L. & King, J. E. (ed.). (1982a). *Primate Behavior.* New York: Academic Press.

Fobes, J. L. & King, J. E. (1982b). Measuring primate learning abilities. In *Primate Behavior*, ed. J. L. Fobes & J. E. King, pp. 289–326. New York: Academic Press.

Fobes, J. L. & King, J. E. (1982c). Vision: the dominant primate modality. In *Primate Behavior*, ed. J. L. Fobes & J. E. King, pp. 219–43. New York: Academic Press.

Foley, R. A. (1987). *Another Unique Species: Patterns In Human Evolutionary Ecology.* New York: John Wiley & Sons.

Foley, R. A. (1988). Hominids, humans and hunter-gatherers: an evolutionary perspective. In *History, Evolution, and Social Change in Hunting and Gathering Societies*, ed. T. Ingold, D. Riches & J. Woodburn pp. 207–221. Oxford: Berg.

Fortes, M. (1983). *Rules and the Emergence of Society.* Royal Anthropological Institute Occasional paper #39.

Foster, M. L. (1983). The symbolic structure of primordial language. In *Human Evolution: Biosocial Perspectives on Human Evolution*, Vol. 4, ed. S. L. Washburn & E. R. McCrown, pp. 77–121. Menlo Park, CA: Benjamin/Cummings.

Fouts, R. S. (1972). The use of guidance in teaching sign language to a chimpanzee. *Journal of Comparative Physiology & Psychology*, **80**, 515–22.

Fox, R. (1979). Kinship categories as natural categories. In *Evolutionary Biology and Human Social Behavior*, ed. N. Chagnon & W. Irons. S. Scituate, MA: Duxbury Press.

Fox, R. (1980). *The Red Lamp of Incest*. London: Hutchinson.

Fredrickson, W. T. & Sackett, G. P. (1984). Kin preferences in primates (*Macaca nemestrina*): relatedness or familiarity? *Journal of Comparative Psychology*, **98**, 29–34.

Frisch, K. von. (1954). *The Dancing Bees: An Account of the Life and Senses of the Honey Bee*, trans. by Dora Ilse. London: Methuen.

Frisch, K. von. (1967). *The Dance Language and Orientation of Bees*, trans. by L. E. Chadwick. Cambridge, Massachusetts: Harvard University Press.

Frumhoff, P. C. & Baker, J. (1988). A genetic component to division of labour within honey bee colonies. *Nature*, **333**, 358–61.

Gallup, G. G., Jr. (1977). Self-recognition in primates. *American Psychologist*, **32**, 329–38.

Gallup, G. G., Jr. (1987). Self-awareness. In *Comparative Primate Biology, Vol. 2B: Behavior, Cognition, and Motivation*, ed. G. Mitchell & J. Erwin, pp. 3–16. New York: Alan R. Liss.

Gamble, C. (1980). Information in the Paleolithic. *Nature*, **283**, 522–3.

Gamble, C. (1982). Interaction and alliance in paleolithic society. *Man* (N.S.), **17**, 92–107.

Gamble, C. (1986). *The Palaeolithic Settlement of Europe*. Cambridge: Cambridge University Press.

Gardner, H. (1985). *The Mind's New Science. A History of the Cognitive Revolution*. New York: Basic Books.

Gardner, R. A. & Gardner, B. T. (1969). Teaching sign language to chimpanzees. *Science*, **165**, 664–672.

Gardner, R. A. & Gardner, B. T. (1973). Teaching sign language to the chimpanzee, Washoe. (16mm. sound film.) State College, Pennsylvania: Psychological Cinema Register.

Gartlan, J. S. (1968). Structure and function in primate society. *Folia Primatologica*, **8**(2), 89–120.

Garton, A. & Pratt, C. (1989). *Learning to Be Literate*. Oxford: Blackwell.

Geertz, C. (1983). *Local Knowledge: Further Essays in Interpretive Anthropology*. New York: Basic Books.

Geist, V. (1971). *Mountain Sheep: A Study in Behavior and Evolution*. Chicago: University of Chicago Press.

Geist, V. (1978). *Life Strategies, Human Evolution, Environmental Design*. New York: Springer Verlag.

Ghiglieri, M. P. (1984). *The Chimpanzees of Kibale Forest: A Field Study of Ecology and Social Structure*. New York: Columbia University Press.

Gibson, J. J. (1979). *The Ecological Approach to Visual Perception*. Boston: Houghton Mifflin.

Gilmore, H. A. (1981). From Radcliffe-Brown to sociobiology: some aspects of the rise of primatology within physical anthropology. *American Journal of Physical Anthropology*, **56**, 387–92.

Goffman, E. (1959). *The Presentation of the Self In Everyday Life*. Garden City, New York: Doubleday.

Goffman, E. (1974). *Frame Analysis: An Essay On the Organization of Experience*. New York: Harper & Row.

Goodall, J. (1965). Chimpanzees of the Gombe Stream Reserve. In *Primate Behavior*, ed. I. DeVore pp. 425–47. New York: Holt, Rinehart & Winston.

Goodall, J. (1967). *My Friends, the Wild Chimpanzees*. Washington, DC: National Geographic Society.

Goodall, J. (1968). Expressive movements and communication in free-ranging chimpanzees: a preliminary report. In *Primates: Studies in Adaptation and Variability*, ed. P. Jay, pp. 313–74. New York: Holt, Rinehart & Winston.

Goodall, J. (1971). *In the Shadow of Man*. Boston: Houghton Mifflin.

Goodall, J. (1986a). *The Chimpanzees of Gombe: Patterns of Behavior*. Cambridge, Massachusetts: Harvard University Press.

Goodall, J. (1986b). *In the Shadow of Man* (Revised edn). Boston, Massachusetts: Houghton Mifflin.

Goodall, J. (1986c). Social rejection, exclusion, and shunning among the Gombe chimpanzees. *Ethology & Sociobiology*, **7**, 227–36.

Goody, J. R. (1956). *The Social Organization of the Lowiili*. London: H. M. Stationery Office.

Gordon, B. (1990). Human language. In *Neurobiology of Comparative Cognition*, ed. R. P. Kesner & D. S. Olton, pp. 21–49. Hillsdale, New Jersey: Lawrence Erlbaum Associates.

Gould, J. L. & Gould, C. G. (1986). Invertebrate intelligence. In *Animal Intelligence: Insights Into the Animal Mind*, ed. R. J. Hoage & L. Goldman, pp. 21–36. Washington, DC: Smithsonian Institution Press.

Gould, S. J. (1980). Is a new and general theory of evolution emerging? *Paleobiology*, **6**, 119–30.

Gould, S. J. (1989). *Wonderful Life: The Burgess Shale and the Nature of History*. New York: W. W. Norton.

Gould, S. J. & Vrba, E. S. (1982). Exaptation – missing term in the science of form. *Paleobiology*, **8**, 4–15.

Gouzoules, S. (1984). Primate mating systems, kin associations, and cooperative behavior: evidence for kin recognition. *Yearbook of Physical Anthropology*, **27**, 99–134.

Gouzoules, S., Gouzoules, H. & Marler, P. (1984). Rhesus monkey (*Macaca mulatta*) scream vocalizations: representational signalling in the recruitment of agonistic aid. *Animal Behavior*, **32**, 182–93.

Grafen, A. (1990). Do animals really recognize kin? *Animal Behavior*, **39**, 42–54.

Grafen, A. (1991a). A reply to Byers & Berkoff. *Animal Behavior*, **41**, 1091–92.

Grafen, A. (1991b). Kin vision?: A reply to Stuart. *Animal Behavior*, **41**, 1095–96.

Grand, T. I. (1984). Motion economy within the canopy: four strategies for mobility. In *Adaptations for Foraging in Nonhuman Primates: Contributions to an Organismal Biology of Prosimians, Monkeys, and Apes*, ed. P. S. Rodman & J. G. H. Cant, pp. 54–72. New York: Columbia University Press.

Graubard, S. R. (1988). *The Artificial Intelligence Debate: False Starts, Real Foundations*. Cambridge, Massachusetts: MIT Press.

Griffin, D. R. (1976). *The Question of Animal Awareness: Evolutionary Continuity of Mental Experience*. New York: Rockefeller University Press.

Griffin, D. R. (1978). Prospects for a cognitive ethology. *Behavioral and Brain Sciences*, **1**(4), 527–38 (With *BBS* commentary. Additional commentary **4**, 615–36, 1980).

Griffin, D. R. (ed.). (1982). *Animal Mind–Human Mind*. Dahlem Life Sciences Workshop, Report 21. Berlin: Springer Verlag.

Griffin, D. R. (1984). *Animal Thinking*. Cambridge, Massachusetts: Harvard University Press.

Grolier, E. de. (1989). Glossogenesis in endolinguistic and exolinguistic perspective: palaeoanthropological data. In *Studies In Language Origins, Vol. 1*, ed. J. Wind, E. G. Pulleyblank, E. de Grolier, & B. H. Bichakjian, pp. 73–138. Amsterdam/Philadelphia: John Benjamins.

Group Processes. (1954–58). See Conference on Group Processes.

Guilford, T. & Dawkins, M. S. (1991). Receiver psychology and the evolution of animal signals. *Animal Behaviour*, **42**, 1–14.

Györi, G. (1990). Historical motivation in the linguistic sign and its cognitive origin. Paper read at the 6th meeting of Language Origins Society.

Hagerstrand, T. (1973). The domain of human geography. In *Directions In Geography*, ed. R. J. Chorley, pp. 67–87. London: Methuen.

Hall, K. R. L. & DeVore, I. (1965). Baboon social behavior. In *Primate Behavior: Field Studies of Monkeys and Apes*, ed. I. DeVore, pp. 53–110. New York: Holt, Rinehart, & Winston.

Halliday, T. R. (1983). The study of mate choice. In *Mate Choice*, ed. P. Bateson, pp. 3–32. Cambridge: Cambridge University Press.

Hamburg, D. A. (1968). Evolution of emotional responses: evidence from recent research. In *Science and Psychoanalysis*, Vol. XII, ed. J. H. Masserman, pp. 30–52. New York: Grune & Stratton.

Hamilton, W. D. (1964). The genetical theory of social behavior, I, II. *Journal of Theoretical Biology*, **7**, 1–52.

Haraway, D. (1978). Animal sociology and a natural economy of the body politic. I: A political physiology of dominance. II: The past is the contested zone: human nature and theories of production and reproduction in primate behavior studies. *Signs*, **4**(1), 21–36, 37–60.

Haraway, D. (1989). *Primate Visions: Gender, Race, and Nature In the World of Modern Science*. New York: Routledge.

Harcourt, A. H. (1988). Alliances in contests and social intelligence. In *Machiavellian Intelligence: Social Expertise and the Evolution of Intellect In Monkeys, Apes, and Humans*, ed. R. Byrne & A. Whiten, pp. 132–152. Oxford: Clarendon Press.

Harcourt, A. H. (1989). Social influences on competitive ability: alliances and their consequences. In *Comparative Socioecology: The Behavioural Ecology of Humans and Other Mammals*, ed. V. Standen & R. A. Foley, pp.223–42. Oxford: Blackwell Scientific Publications.

Hardy, G. H. (1908) Mendelian proportions in a mixed population. *Science*, **28**, 49–50.

Harlow, H. F. (1944). Studies in discrimination learning by monkeys: I. The learning of discrimination series and the reversal of discrimination series. *Journal of General Psychology*, **30**, 3–12.

Harré, R. (1979). *Social Being*. Oxford: Blackwell.

Harré, R. (1984). Vocabularies and theories. In *The Meaning of Primate Signals*, ed. R. Harré & V. Reynolds, pp. 90–110. Cambridge: Cambridge University Press.

Harré, R. & Lamb, R. (eds.). (1986). *The Dictionary of Ethology and Animal Learning*. Cambridge, Massachusetts: MIT Press.

Harré, R. & Reynolds, V. (eds.). (1984). *The Meaning of Primate Signals*. Cambridge: Cambridge University Press.

Harrison, G. A. (1988). Social heterogeneity and biological variation. *Man*, **23**(4), 740–56.

Haugeland, J. (1985). *Artificial Intelligence: The Very IDEA*. Cambridge, Massachusetts: MIT Press.

Hausfater, G. & Hrdy, S. B. (eds.). (1984). *Infanticide: Comparative and Evolutionary Perspectives*. Hawthorne, New York: Aldine.

Heider, F. (1958). *The Psychology of Interpersonal Relations*. New York: Wiley.

Hemelrijk, C. K. & Ek, A. (1990). Reciprocity and interchange of grooming and 'support' in captive chimpanzees. *Animal Behaviour*, **41**, 923–35.

Hewes, G. W. (1973a). An explicit formulation of the relation between tool-using and early language emergence. *Visible Language*, **7**(2), 101–27.

Hewes, G. W. (1973b). Primate communication and the gestural origin of language. *Current Anthropology*, **14**, 5–24.

Hewes, G. W. (1976). The current status of the gestural theory of language origin. In *The Origins & Evolution of Language*. ed. S. Harnad, H. Steklis & J. Lancaster, pp. 482–504. *Annals of the New York Academy of Sciences*, Vol. 280.

Hewes, G. W. (1977). A model for language evolution. *Sign Language Studies*, **15**, 97–168.

Hewes, G. W. (1978). Visual learning, thinking, and communication in human biosocial evolution. In *Visual Learning, Thinking, and Communication*, ed. B. S. Randhana & W. E. Coffman, pp. 1–19. New York: Academic Press.

Hewes, G. W. (1983). The invention of phonemically-based language. In *Glossogenesis: The Origin and Evolution of Language*, ed. E. de Grolier, pp. 143–162. New York: Harwood.

Hewes, G. W. (1989). The upper paleolithic expansion of supernaturalism and the advent of fully developed spoken language. In *Studies In Language Origins, Vol. 1*, ed. J. Wind, E. G. Pulleyblank, E. de Grolier & B. H. Bichakjian, pp. 139–57. Amsterdam/Philadelphia: John Benjamins.

Hewes, G. W. (1993). Review of gestural origin theories. In *Handbook of Human Symbolic Evolution*. ed. A. Lock & C. Peters, Oxford: Oxford University Press.

Hiatt, L. (1968). Gidjingali marriage arrangements. In *Man the Hunter*, ed. R. B. Lee and I. DeVore, pp. 165–75. Chicago: Aldine.

Higgins, E. T. & Bargh, J. A. (1987). Social cognition and social perception. *Annual Review of Psychology*, **38**, 369–425.

Hinde, R. A. (1976). Interactions, relationships and social structure. *Man*, **11**, 1–17.

Hinde, R. A. (1982). *Ethology, its Nature and Relation with Other Sciences.* Oxford: Oxford University Press.

Hinde, R. A. (1983a). A conceptual framework. In *Primate Social Relationships, An Integrated Approach*, ed. R. A. Hinde, pp. 1–7. Oxford: Blackwell.

Hinde, R. A. (1983b). The human species. In *Primate Social Relationships, An Integrated Approach*, ed. R. A. Hinde, pp. 334–339. Oxford: Blackwell.

Hinde, R. A., (ed.). (1983c). *Primate Social Relationships: An Integrated Approach.* Oxford: Blackwell.

Hinde, R. A. (1987). *Individuals, Relationships & Culture: Links Between Ethology & the Social Sciences.* New York: Cambridge University Press.

Hinde, R. A. & Stevenson-Hinde, J. (1976). Towards understanding relationships: dynamic stability. In *Growing Points in Ethology*, ed. P. P. G. Bateson & R. A. Hinde, pp. 451–479. Cambridge: Cambridge University Press.

Hinde, R. A., Perret-Clermont, A.-N. & Stevenson-Hinde, J. (1983). *Social Relationships and Cognitive Development.* Oxford: Clarendon Press.

Hladik, C. M., Charles-Dominique, P. & Petter, J. J. (1980). Feeding strategies of five nocturnal prosimians in the dry forest of the west coast of Madagascar. In *Nocturnal Malagasy Primates*, ed. P. Charles-Dominique, *et al.*, pp. 41–74. New York: Academic Press.

Hoage, R. J. & Goldman, R. (eds.). (1986). *Animal Intelligence: Insights Into the Animal Mind.* Washington, DC: Smithsonian Institution Press.

Hockett, C. F. (1960). Logical considerations in the study of animal communication. In *Animal Sounds and Communication*, ed. W. E. Lanyon & W. N. Tavolga, pp. 392–430. Washington, DC: American Institute of Biological Sciences.

Hockett, C. F. & Altmann, S. A. (1968). A note on design features. In *Animal Communication*, ed. T. Sebeok, pp. 61–72. Bloomington: Indiana University Press.

Holloway, R. L. (ed.). (1974). *Primate Aggression, Territoriality and Xenophobia.* New York: Academic Press.

Holmes, W. G. & Sherman, P. W. (1982). The ontogeny of kin recognition in two species of ground squirrels. *American Zoologist*, **22**, 491-517.

Holmes, W. G. & Sherman, P. W. (1983). Kin recognition in animals. *American Scientist*, **71**(1), 46–55.

Homans, G. C. (1958). Social behavior as exchange. *American Journal of Sociology*, **63**, 597–606.

Homans, G. C. (1961). *Social Behavior: Its Elementary Forms.* New York: Harcourt, Brace & World.

Hrdy, S. B. (1977a). Infanticide as a primate reproductive strategy. *American Scientist*, **65**, 40–9.

Hrdy, S. B. (1977b). *The Langurs of Abu: Female and Male Strategies of Reproduction.* Cambridge, Massachusetts: Harvard University Press.

Huffman, M. A. (1984). Stone-play of *Macaca fuscata* in Arashiyama B troop: transmission of a non-adaptive behavior. *Journal of Human Evolution*, **13**, 725–35.

Huffman, M. A. (1987). Consort intrusion and female mate choice in Japanese macaques. *Ethology*, **75**, 221–34.

Huffman, M. A. (1991). Mate selection and partner preferences in female Japanese macaques. In *The Monkeys of Arashiyama. Thirty-five Years of Research in Japan and the West*, ed. L. M. Fedigan & P. J. Asquith, pp. 101–22.

Huffman, M. A. & Quiatt, D. (1986). Stone handling by Japanese Macaques (*Macaca fuscata*). *Primates*, **27**, 413–23.

Huffman, M. A. & Seifu, M. (1989). Observations on illness and the consumption of a possibly medicinal plant *Vernonia amygdalina* (Del.), by a wild chimpanzee in the Mahale Mountains National Park, Tanzania. *Primates*, **30**(1), 17–25.

Hughes, A. L. (1988). *Evolution and Human Kinship*. Oxford: Oxford University Press.

Humphrey, N. K. (1976). The social function of intellect. In *Growing Points in Ethology*, ed. P. P. G. Bateson & R. A. Hinde, pp. 303–17. Cambridge: Cambridge University Press.

Hurford, J. R. (1989). Biological evolution of the Saussurean sign as a component of the language acquisition device. *Lingua*, **77**, 187–222.

Hutchinson, G. E. (1965). *The Ecological Theater and the Evolutionary Play*. New Haven: Yale University Press.

Imanishi, K. (1960). Social organization of the subhuman primates in their natural habitat. *Current Anthropology*, **1**, 393–407.

Ingold, T. (1990). An anthropologist looks at biology. *Man*, **25**, 208–29.

Isaac, G. (1978a). Food sharing and human evolution: archaeological evidence from the plio-pleistocene of East Africa. *Journal of Anthropological Research*, **34**, 311–25.

Isaac, G. (1978b). The food-sharing behavior of protohuman hominids. *Scientific American*, **238**(4), 90–108.

Isaac, G. (1983). Bones in contention: competing explanations for the juxtaposition of early pleistocene artifacts and faunal remains. In *Animals and Archaeology*, Vol. 1, ed. J. Clutton-Brock & C. Grigson, pp. 3–19. British Archaeology Report.

Jarvis, J. U. M. (1981). Eusociality in a mammal: Cooperative breeding in naked mole-rat colonies. *Science*, **212**, 571–573.

Jay, P. C. (1962). Aspects of maternal behavior among langurs. *Annals of the New York Academy of Sciences*, **102**, 468–76.

Jay, P. C. (1965). The common langur of north India. In *Primate Behavior: Field Studies of Monkeys and Apes*, ed. I. DeVore, pp. 197–249. New York: Holt, Rinehart, & Winston.

Jolly, A. (1966). Lemur and social behaviour and primate intelligence. *Science*, **153**, 501–6.

Jolly, A. (1985). *The Evolution of Primate Behavior*. 2nd edn. New York: MacMillan.

Jolly, C. J. & Plog, F. (1979). *Physical Anthropology and Archeology*, 2nd edn. New York: Knopf.

Jürgens, U. (1990). Vocal communication in primates. In *Neurobiology of Comparative Cognition*, ed. R. P. Kesner & D. S. Olton, pp. 51–76. Hillsdale, New Jersey: Lawrence Erlbaum Associates.

References 303

Kahneman, D. & Tversky, A. (1973). On the psychology of prediction. *Psychological Review*, **80**, 237–51.

Kaplan, J. R. (1977). Patterns of fight interference in free-ranging rhesus monkeys. *American Journal of Physical Anthropology*, **47**, 279–88.

Kaplan, J. R. (1978). Fight interference and altruism in rhesus monkeys. *American Journal of Physical Anthropolgy*, **149**, 449–56.

Kawai, M. (1958). (English transl. 1965). On the system of social ranks in a natural troop of Japanese monkeys. In *Japanese Monkeys*. ed. S. Altmann, University of Alberta.

Kawai, M. (1965). Newly acquired pre-cultural behavior of the natural troop of Japanese monkeys on Koshima Islet. *Primates*, **6**, 1–30.

Kawamura, S. (1959). The process of sub-human culture propagation among Japanese macaques. *Primates*, **2**, 43–60.

Keverne, E. B. (1980). Olfaction in the behaviour of non-human primates. *Symposia of the Zoological Society of London*, **45**, 313–27.

Khanna, S. M. & Tonndorf, J. (1978). Physical and physiological principles controlling auditory sensitivity in primates. In *Sensory Systems of Primates*, ed. C. R. Noback, pp. 23–52. New York: Plenum Press.

King, J. E. & Fobes, J. L. (1982). Complex learning by primates. In *Primate Behavior*, ed. J. L. Fobes & J. E. King, pp. 327–60. New York: Academic Press.

Kinzey, W. G. (1987). *The Evolution of Human Behavior: Primate Models*. Albany: SUNY Press.

Koford, C. (1963a). Group relations in an island colony of rhesus monkeys. In *Primate Social Behavior*, ed. C. H. Southwick, pp. 136–52. Princeton: Van Nostrand.

Koford, C. (1963b). Rank of mothers and sons in bands of rhesus monkeys. *Science*, **141**, 356–57.

Kohler, W. (1925). *The Mentality of Apes*. New York: Harcourt, Brace & Co.

Krebs, J. R. & Dawkins, R. (1984). Animal signals and manipulation. In *Behavioural Ecology, An Evolutionary Approach*, ed. J. R. Krebs & N. B. Davies, pp. 380–402. Oxford: Blackwell Scientific Publications.

Kroeber, A. L. & Kluckhohn, C. (1952). Culture: a critical concept of concepts and definitions. *Harvard University Papers of the Peabody Museum of American Archaeology and Ethnology*, Vol. 47.

Kummer, H. (1968). *Social Organization of Hamadryas Baboons*. Chicago: University of Chicago Press.

Kummer, H. (1971). *Primate Societies: Group Techniques of Ecological Adaptation*. New York: Aldine.

Kummer, H. (1982). Social knowledge in free-ranging primates. In *Animal Mind – Human Mind*, ed. D. R. Griffin, pp. 113–30. Dahlem Conference. Berlin: Springer-Verlag.

Kummer, H., Dasser, V. & Hoyningen-Huene, P. (1990). Exploring primate social cognition: some critical remarks. *Behaviour*, **112**(1–2), 84–98.

Kurland, J. A. (1977). *Kin selection in the Japanese monkey. Contributions to Primatology*, Vol. 12. Basel: S. Karger.

Lack, D. (1954). *The Natural Regulation of Animal Numbers*. Oxford: Oxford University Press.

Laird, J. E. (1988). Recovery from incorrect knowledge in Soar. *Proceedings of the AAAI '88: National Conference on Artificial Intelligence*. Menlo Park, CA: American Association for Artificial Intelligence.

Laird, J., Rosenblum, P. & Newell, A. (1986). *Universal Subgoaling and Learning of Goal Hierarchies*. Boston: Kluwer.

Laitman, J. T. (1983). The evolution of the hominid upper respiratory system and implications for the origins of speech. In *Glossogenetics: The Origin and Evolution of Language*, ed. E. de Grolier, pp. 63–90. New York: Harwood.

Laitman, J. T. (1985). Evolution of the hominid upper respiratory tract: the fossil evidence. In *Hominid Evolution: Past, Present and Future*, ed. P. V. Tobias, pp. 281–6. New York: Alan R. Liss.

Laitman, J. T., Gannon, P. J. & Reidenberg, J. S. (1989). Charting changes in the hominid vocal tract: the fossil evidence. *American Journal of Physical Anthropology*, **78**, 257–58. Abstract only.

Lancaster, J. B. (1984). Introduction. In *Female Primates: Studies By Women Primatologists*, ed. M. F. Small, pp. 1–10. New York: Alan R. Liss.

Latour, B. (1987). *Science in Action: How to Follow Scientists and Engineers Through Society*. Cambridge, Massachusetts: Harvard University Press.

Layton, R. H. (1989). Are sociobiology and social anthropology incompatible? The significance of sociocultural resources in human evolution. In *Comparative Socioecology, The Behavioural Ecology of Humans and Other Mammals*, ed. V. Standen & R. A. Foley, pp. 433–55. Oxford: Blackwell Scientific Publications.

Leach, E. R. (1977). *Rethinking Anthropology*. London: Athlone Press.

Le Boeuf, B. J. (1974). Male–male competition and reproductive success in elephant seals. *American Zoologist*, **14**,163–76.

Leigh, S. R. (1988). Comparisons of rates of evolutionary change in cranial capacity in *Homo erectus* and early *Homo sapiens*. *American Journal of Physical Anthropology*, **75**(2), 237–38. Abstract only.

Levi-Strauss, C. (1949). *The Elementary Structures of Kinship*. Transl. J. H. Bell, J. R. von Sturmer & R. Needham (eds). London: Eyre and Spottiswoode.

Levi-Strauss, C. (1969). *The Elementary Structures of Kinship*. Transl. J. H. Bell, J. R. von Sturmer & R. Needham (eds). London: Eyre & Spottiswoode.

Lieberman, P. (1976). Interactive models for evolution: neural mechanisms, anatomy and behavior. *Annals of the New York Academy of Sciences*, **280**, 660–72.

Lieberman, P. (1983). On the nature and evolution of the biological bases of language. In *Glossogenetics: The Origin and Evolution of Language*, ed. E. de Grolier, pp. 91–114. New York: Harwood.

Lieberman, P. (1984). *The Biology and Evolution of Language*. Cambridge, Massachusetts: Harvard University Press.

Lieberman, P. (1985). On the evolution of human syntactic ability. Its preadaptive bases – motor control and speech. *Journal of Human Evolution*, **14**, 657–68.

Lieberman, P. & Crelin, E. S. (1971). On the speech of Neanderthal man. *Linguistic Inquiry*, **11**, 203–22.

Lieberman, P., Crelin, E. S. & Klatt, D. H. (1972). Phonetic ability and related anatomy of the newborn and adult human, Neandertal man, and the chimpanzee. *American Anthropologist*, **74**, 287–307.

Lindauer, M. (1961). *Communication Among Social Bees*. Cambridge, Massachusetts: Harvard University Press.

Livingstone, F. B. (1969). The Founder effect and deleterious genes. *American Journal of Physical Anthropology*, **30**, 55–60.

Lloyd, J. E. (1986). Firefly communication and deception: "Oh what a tangled web." In *Deception: Perspectives on Human and Nonhuman Animal Deceit*, ed. R. W. Mitchell & N. S. Thompson, pp. 113–28. Albany: SUNY Press.

Lockard, J. S. & Paulhus, D. S. (eds). (1988). *Self-deception: An Adaptive Mechanism?* Englewood Cliffs, New Jersey: Prentice-Hall.

Lovejoy, C. O. (1981). The origin of man. *Science*, **211**, 341–50.

Lumsden, C. J. & Wilson, E. O. (1981). *Genes, Mind & Culture*. Cambridge, Massachusetts: Harvard University Press.

Lyons, J. (1988). Origins of language. In *Origins: The Darwin College Lectures*, ed. A. C. Fabian, pp.141–165. Cambridge: Cambridge University Press.

MacArthur, R. H. (1962). Some generalized theorems of natural selection. *Proceedings of the National Academy of Science, U.S.A.* **48**, 1893–97.

MacFarland, D. (1985). *Animal Behavior: Psychobiology, Ethology and Evolution*. Menlo Park, California: Benjamin/Cummings.

Magnussen, M. and Palsson, H. (1960). *Njal's Saga*. (English trans.) Harmondsworth: Penguin.

Maier, S. F. & Jackson, R. L. (1979). Learned helplessness: all of us were right (and wrong): inescapable shock has multiple effects. *The Psychology of Learning and Motivation*, **13**, 155–218.

Maier, S. F. & Seligman, M. E. P. (1976). Learned helplessness: theory and evidence. *Journal of Experimental Psychology*, **105**, 3–46.

Mair, L. (1972). *An Introduction to Social Anthropology*, (2nd. edn). Oxford: Oxford University Press.

Malinowski, B. (1922). *Argonauts of the Western Pacific*. London: Routledge.

Malthus, T. R. (1798). *An Essay on the Principle of Population*. London: Johnson. (Reprinted in Everyman's Library, 1914.)

Mamet, D. (1987). House of Games. Filmhaus film released by Orion Pictures. (Screenplay and direction by Mamet.)

Marler, P. (1973). A comparison of vocalizations of red-tailed monkeys and blue monkeys, *Cercopithecus ascanius* and *C. mitis*, in Uganda. *Zeitschrift für Tierpsychologie*, **33**, 223–247.

Marler, P. (1975). On the origin of speech from animal sounds. In *The Role of Speech in Language*, ed. J. F. Kavanagh & J. E. Cutting, pp. 11–37. Cambridge, Massachusetts: MIT press.

Martin, R. D. (1990). *Primate Origins and Evolution: A Phylogenetic Reconstruction*. Lawrenceville, New Jersey: Princeton University Press.

Mason, W. A. (1968). Scope and potential of primate research. *Science and Psychoanalysis*, **12**, 110–12.

Maurer, D. (1940). *The Big Con*. Indianapolis: Bobbs-Merrill.

Mawby, R. & Mitchell, R. W. (1986). Feints and ruses: an analysis of deception in sports. In *Deception. Perspectives On Human and Nonhuman Deceit*, ed. R. W. Mitchell & N. S. Thompson, pp. 313–22. Albany: SUNY Press.

Maynard Smith, J. (1964). Group selection and kin selection. *Nature*, **201**, 1145–47.

Maynard Smith, J. (1974). The theory of games and the evolution of animal conflict. *Journal of Theoretical Biology*, **47**, 209–21.

Maynard Smith, J. (1975). *The Theory of Evolution*. Harmondsworth: Penguin Books.

Maynard Smith, J. (1976). Evolution and the theory of games. *American Scientist*, **64**(1), 41–45.

Maynard Smith, J. (1977). Parental investment: a prospective analysis. *Animal Behaviour*, **25**, 1–9.

Maynard Smith, J. (1984). The ecology of sex. In *Behavioural Ecology: An Evolutionary Approach,*, ed. J. R. Krebs & N. B. Davies, pp. 201–21. Oxford: Blackwell Scientific Publications.

McGonigle, B. O. & Chalmers, M. (1977). Are monkeys logical? *Nature*, **267**, 694–6.

McKenna, J. J. (1979). The evolution of allomothering behavior among colobine monkeys: function and opportunism in evolution. *American Anthropologist*, **81**, 818–40.

McKenna, J. J. (1981). Primate infant care-giving behavior: origins, consequences and variability with emphasis on the common Indian langur. In *Parental Care in Mammals*, ed. D. Gubernick & P. Klopfer, pp. 389–417. London: Plenum Press.

McKenna, J. J. (1982). Primate field studies: the evolution of behavior and its socioecology. In *Primate Behavior*, ed. J. L. Fobes and J. E. King, pp. 53–83. New York: Academic Press.

McKenna, J. J. (1987). Parental supplements and surrogates among primates: cross-species and cross-cultural comparisons. In *Parenting Across the Life Span: Biosocial Dimensions*, ed. J. Lancaster, J. Altmann, & A. Rossi, pp. 143–184. New York: Aldine De Gruyter.

McMillan, C. & Duggleby, C. (1981). Interlineage genetic differentiation among rhesus macaques on Cayo Santiago. *American Journal of Physical Anthropology*, **56**, 305–12.

Mead, G. H. (1934). *Mind, Self and Society From the Standpoint of a Social Behaviorist*. Chicago: University of Chicago Press.

Meikle, D. B. & Vessey, S. H. (1981). Nepotism among rhesus monkey brothers. *Nature*, **94**, 160–1.

Menzel, C. B. (1980). Head-cocking and visual perception in primates. *Animal Behaviour*, **28**, 151–9.

Menzel, E. W., Jr. (1974). A group of young chimpanzees in a one-acre field. In *Behavior of Non-human Primates*, Vol. 5, ed. A. M. Schrier & F. Stollnitz. New York: Academic

Menzel, E. W., Jr., Savage-Rumbaugh, E.S., and Lawson, J. (1985). Chimpanzee (*Pan troglodytes*) spatial problem solving with the use of mirrors and televised equivalents of mirrors. *Journal of Comparative Psychology*, **99**, 211–17.

Miles, H. L. (1986). How can I tell a lie? Apes, language, and the problem of deception. In *Deception: Perspectives on Human and Nonhuman Deceit*, ed. R. W. Mitchell & N. S. Thompson, pp. 245–66. Albany: SUNY Press.

Milo, R. G. & Quiatt, D. (1993a). Evidence for glottogenesis in anatomically modern *Homo sapiens*: some tentative conclusions. In ed. D. Quiatt & J. Itani, *Hominid Culture in Primate Perspective*. Boulder, Colorado: University Press of Colorado. (In press)

Milo, R. G. & Quiatt, D. (1993b). Group selection in the biocultural evolution of language: fate of the Neandertals revisited. In *The Biology of Language: Essentialist versus Evolutionist*. ed. J. Fisiak. Amsterdam: John Benjamins. (In press)

Mischel, W. (1968). *Personality and Assessment*. New York: Wiley.

Mitani, J. C. & Marler, P. (1989). A phonological analysis of male gibbon singing behavior. *Behaviour*, **109**, 20–45.

Mitchell, R. W. (in press 1993). Mental models of mirror self-recognition: two theories. *New Ideas in Psycology*, **11**.

Mitchell, R. W. & Thompson, N. S. (eds). (1986). *Deception: Perspectives on Human and Nonhuman Deceit*. Albany: SUNY Press.

Mitchell, R. W. & Thompson, N. S. (1993). Taking animals seriously. In *Anthropomorphism, Anecdotes, and Animals: The Emperor's New Clothes?*, ed. R.W. Mitchell & N.S. Thompson. Lincoln, Nebraska: University of Nebraska Press. (In press).

Mohnot, S. M. (1971). Some aspects of social changes and infant-killing in the hanuman langur, *Presbytis entellus* (Primates: Cercopithecidae) in western India. *Mammalia*, **35**, 175–98.

Morris, D. (1967). *The Naked Ape*. New York: McGraw-Hill.

Moss, C. (1988). *Elephant Memories: Thirteen Years of Life in an Elephant Family*. New York: William Morrow.

Murdock, G. P. (1949). *Social Structure*. New York: Macmillan.

Napier, J. R. & Napier, P. H. (1985). *The Natural History of the Primates*. Cambridge, Massachusetts: MIT Press.

Newell, A. (1990). *Unified Theories of Cognition. The William James Lectures*. Cambridge, Massachusetts: Harvard University Press.

Newell, A. & Simon, H. A. (1963). GPS: a program that simulates human thought. In *Computers and Thought*, ed. E. A. Feigenbaum & J. Feldman, pp. 279–93. New York: McGraw-Hill.

Nishida, T. (1979). The social structure of chimpanzees of the Mahale mountains. In *The Great Apes*, ed. D. A. Hamburg & E. R. McCown, pp. 73–121. Menlo Park: Benjamin/Cummings.

Odum, E. P. (1969). The attitude lag. *BioScience*, **19**, 403.

Odum, E. P. (1970). Optimum population and environment: a Georgian microcosm. *Current History*, **58**, 355–9; 365.

Odum, E. P. (1971). *Fundamentals of Ecology*, 3rd edn. Philadelphia: W. B. Saunders.

Orians, G. H. (1969). On the evolution of mating systems in birds and mammals. *American Naturalist*, **103**, 589–603.

Packer, C. (1977). Reciprocal altruism in olive baboons. *Nature*, **265**, 441–3.

Parker, G. A. (1974). Assessment strategy and the evolution of fighting behaviour. *Journal of Theoretical Biology*, **47**, 223–43.

Parker, G. A. (1983). Arms races in evolution – an ESS to the opponent – independent costs games. *Journal of Theoretical Biology*, **101**, 619–48.

Parker, G. A. (1984). Evolutionarily stable strategies. In *Behavioural Ecology:*

An Evolutionary Approach, 2nd edn., ed. J. R. Krebs & N. B. Davies, pp. 30–61. Oxford: Blackwell Scientific Publications.

Passingham, R. E. (1982). *The Human Primate*. San Francisco: W. H. Freeman.

Pattee, H. H. (1970). The problem of biological hierarchy. In *Towards A Theoretical Biology*, ed. C. H. Waddington, Vol. 3, pp. 117–36. Edinburgh: Edinburgh University Press.

Patterson, F. (1978a). The gestures of a gorilla: language acquisition in another pongid. *Brain & Language*, **5**, 72–97.

Patterson, F. (1978b). Linguistic capabilities of a young lowland gorilla. In *Sign Language and Language Acquisition in Man and Ape: New Dimensions in Comparative Pedolinguistics*, ed. F. C. Peng, pp. 161–201. Boulder, Colorado: Westview Press.

Patterson, F. (1980a). Innovative uses of language by a gorilla: a case study. In *Children's Language*, Vol. 2, ed. K. Nelson, pp. 497–561. New York: Gardner.

Patterson, F. (1980b). In search of man: experiments in primate communication. *Michigan Quarterly Review*, **19**(1), 95–114.

Patterson, F. (1981). *Conversations With a Gorilla*. New York: Holt, Rinehart, & Winston.

Patterson, F. (1986). The mind of the gorilla: conversation and conservation. In ed. K. Benirschke, *Primates: The Road To Self-sustaining Populations*, pp. 933–47. New York: Springer-Verlag.

Patterson, F. G. & Linden, E. (1981). *The Education of Koko*. New York: Holt, Rinehart & Winston.

Pehrson, R. N. (1969). *The Social Organization of the Marri Baluch*. Chicago: Aldine Publishing Co.

Plotkin, H. C. & Odling-Smee, F. J. (1981). A multiple-level model of evolution and its implications for sociobiology. *Behavioral & Brain Sciences*, **4**, 225–68.

Postman, L., Bruner, J. S., & McGinnies, E. (1948). Personal values as selective factors in perception. *Journal of Abnormal and Social Psychology*, **43**, 142–54.

Potts, R. B. (1983). Foraging for faunal resources by early hominids at Olduvai Gorge, Tanzania. In *Animals and Archaeology, Vol. 1, Hunters and Their Prey*, ed. J. Clutton-Brock & C. Grigson, pp. 51–62. British Archaeological Reports International Series 163.

Premack, D. (1972). Language in chimpanzee? *Science*, **172**, 808–22.

Premack, D. (1976). *Intelligence In Ape and Man*. Hillsdale, New Jersey: Lawrence Erlbaum.

Premack, D. (1984). Pedagogy and aesthetics as sources of culture. In *Handbook of Cognitive Neuroscience*, ed. M. S. Gazzaniga, pp. 15–35. New York: Plenum Press.

Premack, D. (1986). *Gavagai! Or the Future History of the Animal Language Controversy*. Cambridge, Massachusetts: MIT Press.

Pylyshyn, Z. W. (1984). *Computation and Cognition: Toward a Foundation of Cognitive Science*. Cambridge, Massachusetts: MIT Press.

Quiatt, D. (1966). *Social Dynamics of Rhesus Monkey Groups*. Ann Arbor: University Microfilms.

Quiatt, D. (1984). Devious intentions of monkeys & apes? In *The Meaning of Primate Signals*, ed. R. Harré & V. Reynolds, pp. 9–40. Cambridge: Cambridge University Press.

Quiatt, D. (1985). The 'household' in non-human primate evolution: a basic linking concept. *Anthropologea Contemporanea*, **3**, 187–93.

Quiatt, D. (1986a). Japanese macaque stone handlers. Videotape available from the Primate Center Library, Wisconsin Regional Primate Research Center, Madison, WI.

Quiatt, D. (1986b). Juvenile/adolescent role functions in a rhesus monkey troop: an application of household analysis to nonhuman primate social organization. In *Primate Ontogeny, Cognition and Social Behaviour*, ed. J. Else & P. Lee, pp.281–9. Cambridge: Cambridge University Press.

Quiatt, D. (1987a). A-Group: one example of gibbon social organization. Videotape available from the Primate Center Library, Wisconsin Regional Primate Research Center, Madison, WI.

Quiatt, D. (1987b). Aunting, alloparenting, infant handling: evolving views of adaptive function. *Primate Report*, **11**, 67–73.

Quiatt, D. (1988a). Regulation of mating choice in nonhuman primates. In *Human Mating Patterns. Society for the Study of Human Biology Symposium*, ed. C. G. N. Mascie-Taylor & A. J. Boyce, Vol. 28, pp. 132–151. Cambridge: Cambridge University Press.

Quiatt, D. (1988b). Which are more easily deceived, friends or strangers? *Behavioral and Brain Sciences*, **11**, 740–41.

Quiatt, D. (1990). Ape language revisited. *Reviews in Anthropology*, **18**, 247–57.

Quiatt, D. & Everett, J. (1982). How can sperm competition work? *American Journal of Primatology, Supplement*, **1**, 161–69.

Quiatt, D. & Kelso, J. (1985). Household economics and hominid origins. *Current Anthropology*, **26**, 207–22.

Quiatt, D. & Kelso, J. (1987). The concept of the household: linking behavior and genetic analyses. *Human Evolution*, **2**, 429–35.

Quiatt, D., Everett, J., Luerssen, S. & Murdock, G. (1981). Problems in representing behavioral space-time. In *Primate Behavior and Sociobiology*, ed. A. B. Chiarelli & R. S. Corruccini, pp. 121–9. Berlin/New York: Springer-Verlag.

Radcliffe-Brown, A. R. (1952). *Structure and Function in Primitive Society*. London: Cohen and West.

Radcliffe-Brown, A.R. & Forde, D. (1950). *African Systems of Kinship and Marriage*. Oxford University Press.

Reynolds, P. C. (1976). The emergence of early hominid social organization: 1. The attachment systems. *Yearbook of Physical Anthropology*, **20**, 73–95.

Reynolds, P. C. (1981). *On the Evolution of Human Behavior: The Argument From Animals to Man*. Berkeley: University of California Press.

Reynolds, V. (1975). *How Wild are the Gombe Chimpanzees? Man*, **10**, 123–125.

Reynolds, V. (1976). *The Biology of Human Action*. San Francisco: W. H. Freeman.

Reynolds, V. (1984). Social changes in a group of rhesus monkeys. In *The Meaning of Primate Signals*, ed. R. Harré & V. Reynolds, pp. 208–25. Cambridge: Cambridge University Press.

Reynolds, V. (1986). Primate social thinking. In *Primate Ontogeny, Cognition and Social Behaviour*, ed. J. G. Else & P. C. Lee, pp. 53–60. Cambridge University Press.

Reynolds, V. & Reynolds, F. (1965). Chimpanzees of the Budongo Forest. In *Primate Behavior. Field Studies of Monkeys and Apes*, ed. I. DeVore, pp. 368–424. New York: Holt, Rinehart and Winston.

Reynolds, V. & Tanner, R. (1983). *The Biology of Religion*. London: Longman.

Richard, A. F. (1981). Changing assumptions in primate ecology. *American Anthropologist*, **83**, 517–33.

Richard, A. F. (1985). *Primates in Nature*. San Francisco: W. H. Freeman.

Riesen, A. H. (1982). Primate perceptual processes. In *Primate Behavior*, ed. J. L. Fobes & J. E. King, pp. 271–86. New York: Academic Press.

Rindos, D. (1986). The evolution of the capacity for culture: sociobiology, structuralism, and cultural selectionism. *Current Anthropology*, **27**, 315–32.

Robinson, G. E. & Page, R. W. (1988). Genetic determination of guarding and undertaking in honeybee colonies. *Nature*, **333**, 358–61.

Rodman, P. S. (1988). Resources and group sizes of primates. In *The Ecology of Social Behavior*, ed. C. N. Slobodchikoff, pp. 83–108. San Diego: Academic Press.

Rohles, F. H. & Devine, J. V. (1966). Chimpanzee performance on a problem involving the concept of middleness. *Animal Behavior*, **14**, 159–62.

Rohles, F. H. & Devine, J. V. (1967). Further studies of the middleness concept with the chimpanzee. *Animal Behavior*, **15**, 107–12.

Rowell, T. E. (1967). A quantitative comparison of the behavior of a wild and a caged baboon group. *Animal Behaviour*, **15**, 499–589.

Rowell, T. E. (1988). Beyond the one-male group. *Behaviour*, **104**, 189–201.

Rubenstein, D. I. & Wrangham R. W. (eds). (1986). *Ecological Aspects of Social Evolution: Birds and Mammals*. Princeton, New Jersey: Princeton University Press.

Rudel, R. G. & Teuber, H. L. (1964). Cross-modal transfer of shape discrimination by children. *Neuropsychologia*, **2**, 1–8.

Rumbaugh, D. M. (ed.). (1977). *Language Learning By a Chimpanzee: The LANA Project*. New York: Academic Press.

Rumbaugh, D. M. & Gill, T. V. (1976). Lana's mastery of language skills. In *Origin and Evolution of Language and Speech*. ed. H. Steklis, S. Harnad, & J. Lancaster, *Annals of the New York Academy of Sciences*, **280**, 562–78.

Sacks, O. (1987). *The Man Who Mistook His Wife For a Hat and Other Clinical Tales*. New York: Summit Books.

Samuels, A., Silk, J. B. & Altmann, J. (1987). Continuity and change in dominance relations among female baboons. *Animal Behaviour*, **35**, 785–93.

Saussure, F. de. (1959). *Course in General Linguistics*. (transl. W. Baskin). New York: McGraw-Hill.

Savage-Rumbaugh, E. S. (1984a). Verbal behavior at a procedural level in the chimpanzee. *Journal of the Experimental Analysis of Behavior*, **41**, 223–50.

Savage-Rumbaugh, E. S. (1984b). *Pan paniscus* and *Pan troglodytes*. Contrasts in preverbal communicative competence. In *The Pygmy Chimpanzee: Evolutionary Biology and Behavior*, ed. R. L. Susman, pp. 395–414. New York: Plenum Press.

Savage-Rumbaugh, E. S. (1986). *Ape Language: From Conditioned Response to Symbol*. New York: Columbia University Press.

Savage-Rumbaugh, E. S. (1990). Implications of the cognitive and linguistic abilities of the Bonobo for theories of the development of hominid culture. Paper delivered at XIIIth Congress of International Primatological Society, Kyoto, Japan.

Savage-Rumbaugh, S. & McDonald, K. (1988). Deception and social manipulation in symbol-using apes. In *Machiavellian Intelligence: Social Expertise and the Evolution of Intellect in Monkeys, Apes, and Humans*, ed. R. Byrne & A. Whiten, pp. 224–37. Oxford: Clarendon Press.

Schiller, P. H. (1957). Innate motor actions as a basis of learning: manipulation patterns in the chimpanzee. In *Instinctive Behavior: The Development of a Modern Concept*, ed. C.H. Schiller, pp. 264–87. New York: International Universities Press.

Scott, J. P. (1968). *Early Experience and the Organization of Behavior*. Belmont, California: Brooks/Cole Publishing Company.

Selye, H. (1956). *The Stress of Life*. New York: McGraw-Hill.

Seyfarth, R. M. (1984). What the vocalizations of monkeys mean to humans and what they mean to the monkeys themselves. In *The Meaning of Primate Signals*, ed. R. Harré & V. Reynolds, pp. 43–56. Cambridge: University Press.

Seyfarth, R. M. & Cheney, D. L. (1988). Do monkeys understand their relations? In *Machiavellian Intelligence: Social Expertise and the Evolution of Intellect in Monkeys, Apes, and Humans*, ed. R. Byrne & A. Whiten, pp. 69–84. Oxford: Clarendon Press.

Seyfarth, R. M., Cheney, D. L. & Marler, P. (1980a). Vervet monkey alarm calls: semantic communication in a free-ranging primate. *Animal Behaviour*, **28**, 1070–94.

Seyfarth, R. M., Cheney, D. L. & Marler, P. (1980b). Monkey responses to three different alarm calls: evidence for semantic communication and predator classification. *Science*, **210**, 801–3.

Shannon, C. E. (1949). *A Mathematical Theory of Communication*. Urbana: University of Illinois Press.

Sheets-Johnstone, M. (1990). Hominid bipedality and sexual selection theory. In *The Roots of Thinking*, ed. M. Sheets-Johnstone, pp. 167–202. Philadelphia: Temple University Press.

Sherman, P. W. (1981). Kinship, demography, and Belding's ground squirrel nepotism. *Behavioral Ecology and Sociobiology*, **8**, 251–60.

Shipman, P. (1983). Early hominid lifestyle: hunting and gathering or foraging and scavenging? In *Animals and Archaeology*, Vol. 1, ed. J. Clutton-Brock & C. Grigson, pp. 31–49. Oxford: British Archaeological Reports.

Shipman, P. (1984). Scavenger hunt. *Natural History*, **93**(4), 20–7.

Silk, J. B. (1982). Altruism among female Macaca radiata: explanations and analysis of patterns of grooming and coalition formation. *Behaviour*, **79**, 162–88.

Silk, J. B. (1987). Social behavior in evolutionary perspective. In *Primate Societies*, ed. B. B. Smuts, *et al.*, pp. 318–29. Chicago: Chicago University Press.

Simons, E. L. (1972). *Primate Evolution: An Introduction to Man's Place in Nature*. New York: Macmillan.

Slater, P. J. B. (1973). Describing sequences of behavior. In *Perspectives in Ethology*, Vol. 1, ed. P. P. G. Bateson & P. H. Klopfer, pp. 131–53. New York: Plenum Press.

Slobodchikoff, C. N. (1984). Resources and the evolution of social behavior, In *A new Ecology: Novel Approaches to Interactive Systems*, ed. P. W. Price, C. N. Slobodchikoff & W. S. Gaud, pp. 227–51. New York: Wiley.

Small, M. F. (1988). Female primate sexual behavior and conception: are there really sperm to spare? *Current Anthropology*, **29**, 81–100.

Small, M. F. (1990). Alloparental behaviour in Barbary macaques, *Macaca sylvanus*. *Animal Behaviour*, **39**, 297–306.

Smith, D. G. & Small, M. F. (1987). Mate choice by lineage in three captive groups of rhesus macaques (*Macaca mulatta*). *American Journal of Physical Anthropology*, **73**, 185–191.

Smith, E. O. (1987). Deception and evolutionary biology. *Cultural Anthropology*, **2**, 50–64.

Smith, W. J. (1977). The behavior of communication: an ethological approach. Cambridge, Massachusetts: Harvard University Press.

Smith, W. J. (1984). An 'informational' perspective on manipulation. In *Deception: Perspectives on Human and Nonhuman Deceit*, ed. R. W. Mitchell & N. S. Thompson, pp. 71–86. Albany: SUNY Press.

Smuts, B. B. (1985). *Sex and Friendship in Baboons*. New York: Aldine.

Smuts, B. B., Cheney, D., Seyfarth, R., Wrangham, R. & Struhsaker, T. (eds). (1987). *Primate Societies*. Chicago: University of Chicago Press.

Snowdon, C. T. (1986). Vocal communication. In *Comparative Primate Biology, Vol. 2A: Behavior, Conservation, and Ecology*, ed. G. Mitchell & J. Erwin, pp. 495–530. New York: Alan R. Liss.

Southwick, C. H., Beg, M. A. & Siddiqi, M. R. (1965). Rhesus monkeys in North India. In *Primate Behavior: Field Studies of Monkeys and Apes*, ed. I. DeVore, pp. 111–59. New York: Holt, Rinehart and Winston.

Srinivas, M. N. (1962). *Caste in Modern India*. Bombay and London: Asia Publishing House.

Staddon, J. E. R. (1989). *Limits to Action: The Allocation of Individual Behavior*. New York: Academic Press.

Stammbach, E. (1988). An experimental study of social knowledge: adaptation to the special manipulative skills of single individuals in a *Macaca fascicularis* group. In *Machiavellian Intelligence: Social Expertise and the Evolution of Intellect in Monkeys, Apes, and Humans*, ed. R. Byrne & A. Whiten, pp. 309–26. Oxford: Oxford University Press.

Standen, V. & Foley R. A. (eds). (1989a). *Comparative Socioecology. The Behavioural Ecology of Human and Other Mammals*. Oxford: Blackwell Scientific.

Standen, V. & Foley, R. A. (1989b). Social organization: the influence of resources. In *Comparative Socioecology: The Behavioural Ecology of Human and Other Mammals*, ed. V. Standen & R. A. Foley, pp. 125–129. Oxford: Blackwell Scientific.

Stebbins, G. L. & Ayala, F. J. (1981). Is a new evolutionary synthesis necessary? *Science*, **213**, 967–71.

Stein, D.M. (1984). *The Sociobiology of Infant and Adult Male Baboons.* Norwood, NJ: Ablex Publ. Corp.

Stern, C. (1973). *Principles of Human Genetics,* 3rd edn. San Francisco: W. H. Freeman.

Stringer, C. (1984). Human evolution and biological adaptation in the Pleistocene. In *Hominid Evolution and Community Ecology,* ed. R. Foley, pp. 55–83. New York: Academic Press.

Stringer, C. B. & Andrews, P. (1988). Genetic and fossil evidence for the origin of modern humans. *Science,* **239,** 1263–8.

Struhsaker, T. T. (1967). Auditory communication in vervet monkeys. In *Social Communication Among Primates,* ed. S. A. Altmann, pp. 281–324. Chicago: University of Chicago Press.

Struhsaker, T. T. & Leland, L. (1987). Colobines: infanticide by adult males. In *Primate Societies,* ed. B. B. Smuts, pp. 83–97. Chicago: University of Chicago Press.

Strum, S. (1987). *Almost Human: A Journey Into the World of Baboons.* New York: Random House.

Suarez, S. D. & Gallup, G. G. Jr. (1981). Self-recognition in chimpanzees and orangutans but not gorillas. *Journal of Human Evolution,* **10,** 175–88.

Sugiyama, Y. (1965). On the social change of hanuman langurs (*Presbytis entellus*) in their natural conditions. *Primates,* **6,** 213–47.

Susman, R. L. (1989). New hominid fossils from the Swartkrans formation (1979–1986 excavations): postcranial specimens. *American Journal of Physical Anthropology,* **79,** 451–74.

Sussman, G. J. (1975). *A Computer Model of Skill Acquisition.* New York: American Elsevier.

Sussman, R. W. (ed.). (1979). *Primate Ecology: Problem-oriented Field Studies.* New York: John Wiley & Sons.

Symons, D. (1978). *Play and Aggression: A Study of Rhesus Monkeys.* New York: Columbia University Press.

Tanner, N. M. (1981). *On Becoming Human.* Cambridge: Cambridge University Press.

Teleki, G. (1973). *The Predatory Behavior of Chimpanzees.* Lewisburg, Pennsylvania: Bucknell University Press.

Terrace, H. S. (1984). 'Language' in apes. In *The Meaning of Primate Signals,* ed. R. Harré & V. Reynolds, pp. 179–203. Cambridge: Cambridge University Press.

Terrace, H. S., Petitto, L. A., Sanders R. J. & T. G. Bever. (1979). Can an ape create a sentence? *Science,* **206,** 891–900.

Thomas, R. K. (1980). Evolution of intelligence: an approach to its assessment. *Brain, Behavior & Evolution,* **17,** 454–72.

Thomas, R. K. (1986). Vertebrate intelligence: a review of the laboratory research. In *Animal Intelligence: Insights Into the Animal Mind,* ed. R. J. Hoage & L. Goldman, pp. 37–56. Washington, DC: Smithsonian Press.

Thomas, R. K. & Chase, L. (1980). Relative numerousness judgments by squirrel monkeys. *Bulletin of the Psychonomic Society,* **16,** 79–82.

Thomas, R. K., Fowlkes, D., & Vickery, J. D. (1980). Conceptual numerousness, judgments by squirrel monkeys. *American Journal of Psychology*, **93**, 247–57.

Thompson, N. S. (1986). Deception and the concept of behavioral design. In *Deception: Perspectives on Human and Nonhuman Primate Deceit*, ed. R. W. Mitchell and N. S. Thompson, pp. 53–65. Albany: SUNY Press.

Tinbergen, N. (1951). The Study of Instinct. Oxford: Oxford University Press.

Tinbergen, N. (1963). On aims and methods of ethology. *Zeitschrift für Tierpsychologie*, **20**, 410–33.

Tooby, J. & DeVore, I. (1987). The reconstruction of hominid behavioral evolution through strategic modeling. In *The Evolution of Human Behavior: Primate Models*, ed. W. G. Kinzey, pp. 183–237. Albany: SUNY Press.

Trivers, R. L. (1971). The evolution of reciprocal altruism. *Quarterly Review of Biology*, **46**, 35–57.

Trivers, R. L. (1972). Parental investment and sexual selection. In *Sexual Selection and the Descent of Man, 1871–1971*, ed. B. Campbell, pp. 136–79. Chicago: Aldine.

Trivers, R. L. (1974). Parent–offspring conflict. *American Zoologist*, **14**, 249–64.

Trivers, R. L. (1985). *Social Evolution*. Menlo Park, California: Benjamin/ Cummings.

Trivers, R. L. & Willard, D. E. (1973). Natural selection of parental ability to vary the sex ratio of offspring. *Science*, **179**, 90–2.

Tutin, C. E. G. (1980). Reproductive behavior of wild chimpanzees in the Gombe National Park, Tanzania. *Journal of Reproduction and Fertility, Supplement*, **28**, 43–57.

Vasek, M. E. (1986). Lying as a skill: the development of deception in children. In *Deception: Perspectives in Human and Nonhuman Deceit*, ed. R. W. Mitchell & N. S. Thompson, pp. 271–92. Albany: SUNY Press.

Vogel, C. & Loch, H. (1984). Reproductive parameters, adult-male replacements, and infanticide among free-ranging langurs (*Presbytis entellus*) at Jodhpur (Rajasthan), India. In *Infanticide: Comparative and Evolutionary Perspectives*, ed. G. Hausfater & S. B. Hrdy, pp. 237–55. Hawthorne, New York: Aldine.

Waal, F. B. M. de. (1982). *Chimpanzee Politics*. London: Jonathan Cape.

Waal, F. B. M. de. (1984). Sex-differences in the formation of coalitions among chimpanzees. *Ethology & Sociobiology*, **5**, 239–55.

Waal, F. B. M. de. (1986). Deception in the natural communication of chimpanzees. In *Deception: Perspectives On Human and Nonhuman Deceit*, ed. R. W. Mitchell & N. S. Thompson, pp. 271–92. Albany: SUNY Press.

Waal, F. B. M. de. (1989a). Dominance 'style' and primate social organization. In *Comparative Socioecology: The Behavioural Ecology of Humans and Other Mammals*, ed. V. Standen & R. A. Foley, pp. 243–63. Oxford: Blackwell Scientific.

Waal, F. B. M. de (1989b). Food sharing and reciprocal obligations among chimpanzees. *Journal of Human Evolution*, **18**, 433–59.

Waal, F. B. M. de. (1989c). *Peace-making Among Primates*. Cambridge, Massachusetts: Harvard University Press.

Waal, F. B. M. de & van Roosmalen, J. (1979). Reconciliation and consolation among chimpanzees. *Behavioral Ecology and Sociobiology*, **5**, 55–66.

Waldrop, M. M. (1988a). Toward a unified theory of cognition. *Science*, **241**, 27–9.

Waldrop, M. M. (1988b). Soar: a unified theory of cognition? *Science*, **241**, 296–8.

Waser, P. (1977). Individual recognition, intragroup cohesion, and intergroup spacing: evidence from sound playback to forest monkeys. *Behaviour*, **10**, 28–74.

Waser, P. (1982a). Polyspecific associations: do they occur by chance? *Animal Behaviour*, **30**, 1–8.

Waser, P. (1982b). The evolution of male loud calls among mangabeys and baboons. In *Primate Communication*, ed. C. T. Snowdon, C. H. Brown & M. R. Petersen, pp. 117–43. Cambridge: Cambridge University Press

Waser, P. (1984). 'Chance' and mixed-species associations. *Behavioral Ecology and Sociobiology*, **15**, 197–202.

Washburn, S. L. & Lancaster, C. S. (1968). The evolution of hunting. In *Man the Hunter*, ed. R. B. Lee & I. DeVore, pp. 293–303. Chicago: Aldine.

Watson, J. D. & Crick, F. C. (1953). Genetical implication of the structure of deoxyribose nucleic acid. *Nature*, **171**, 964–7.

Wescott, R. W. (ed.). (1974). *Language origins*. Mintwood, MD: Linstock.

Wheeler, P. E. (1985). The evolution of bipedality and loss of functional body hair in hominids. In *Essays in Human Sociobiology*, ed. J. Wind, pp. 91–8. London: Academic Press. (Orig. in *Journal of Human Evolution*, **13**, 91–8, 1984).

Whiten, A. & Byrne, R. W. (1986). The St. Andrews catalogue of tactical deception in primates. *St. Andrews Psychological Reports No. 10.*

Whiten, A. & Byrne, R. W. (1988a). The manipulation of attention in primate tactical deception. In *Machiavellian Intelligence: Social Expertise and the Evolution of Intellect in Monkeys, Apes and Humans*, ed. R. W. Byrne & A. Whiten, pp. 211–23. Oxford: Oxford University Press.

Whiten, A. & Byrne, R. W. (1988b). Tactical deception in primates. *Behavioral and Brain Sciences*, **11**(2), 233–73.

Wickler, W. (1967). Socio-sexual signals and their intra-specific imitation among primates. In *Primate Ethology*, ed. D. Morris, pp. 69–147. Chicago: Aldine.

Wickler, W. (1968). *Mimicry In Plants and Animals*. New York: McGraw-Hill.

Williams, G. C. (1966). *Adaptation and Natural Selection*. Princeton, New Jersey: Princeton University Press.

Wilson, E. O. (1975). *Sociobiology: The New Synthesis*. Cambridge, Massachusetts: Harvard University Press.

Wilson, P. J. W. (1983). *Man the Promising Primate*. New Haven: Yale University Press.

Wind, J. (1989). The evolutionary history of the human speech organs. In *Studies in Language Origins*, Vol. 1, ed. J. Wind, E. G. Pulleyblank, E. de Grolier & B. H. Bichakjian, pp. 173–197. Amsterdam: John Benjamins.

Winograd, T. (1973). A procedural model of language understanding. In *Computer Models of Thought and Language*, ed. R. C. Schank & K. M. Colby, pp. 152–186. San Francisco: W. H. Freeman.

Wobst, M. (1976). Locational relationships in paleolithic society. *Journal of Human Evolution*, 5, 49–58.

Wobst, M. (1977). Stylistic behavior and information exchange. In *For the Director: Essays in Honor of James B. Griffin*, ed. C. Cleland. *Ann Arbor: Anthropology Papers of the University of Michigan*, 61, 317–42.

Woodburn, J. (1968). Stability and flexibility in Hadza residential groupings. In *Man the Hunter*, ed. R. B. Lee and I. DeVore, pp. 103–10. Chicago: Aldine.

Wrangham, R. W. (1974). Artificial feeding of chimpanzees and baboons in their natural habitat. *Animal Behaviour*, 22, 83–93.

Wrangham, R. W. (1977). Feeding behavior of chimpanzees in Gombe National Park, Tanzania. In *Primate Ecology*, ed. T. H. Clutton-Brock, pp. 503–38. New York: Academic Press.

Wrangham, R. W. (1980). An ecological model of female bonded primate groups. *Behaviour*, 75, 262–300.

Wrangham, R. W. (1988). Bridging the gaps: social relationships in animals and people. *Rackham reports, 1987*. pp. 20–39 Horace H. Rackham School of Graduate Studies, The University of Michigan.

Wrangham, R. W. & Goodall, J. (1989). Chimpanzee use of medicinal leaves. In *Understanding Chimpanzees*, ed. P. G. Helne & L. A. Marquandt, pp. 22–37. Cambridge, Massachusetts: Harvard University Press.

Wrangham, R. W. & Nishida, T. (1983). *Aspilia* spp. leaves: a puzzle in the feeding behavior of wild chimpanzees. *Primates*, 24(2), 276–82.

Wynn, T. (1979). The intelligence of later Acheulean hominids. *Man*, 14, 371–91.

Wynn, T. (1981). The intelligence of Oldowan hominids. *Journal of Human Evolution*, 10, 529–41.

Wynn, T. (1985). Piaget, stone tools, and the evolution of human intelligence. *World Archaeology*, 17, 32–43.

Wynn, T. (1989a). Tools and the evolution of human intelligence. In *Machiavellian Intelligence, Social Expertise, and the Evolution of Intellect in Monkeys, Apes, and Humans*, ed. R. Byrne & A. Whiten, pp. 271–84. Oxford: Clarendon Press.

Wynn, T. (1989b). The evolution of spatial competence. *University of Illinois Studies in Archaeology No. 17*. Urbana: University of Illinois Press.

Wynne-Edwards, V. C. (1962). *Animal Dispersion in Relation to Social Behaviour*. Edinburgh: Oliver and Boyd.

Wynne-Edwards, V. C. (1964). A reply to Maynard Smith's rejoinder. *Nature*, 201, 1147.

Young, M. & Harlow, H. F. (1943a). Solution by rhesus monkeys of a problem involving the Weigl-principle using the oddity method. *Journal of Comparative Psychology*, 35, 205–17.

Young, M. & Harlow, H. F. (1943b). Generalization by rhesus monkeys of a problem involving the Weigl-principle using the oddity method. *Journal of Comparative Psychology*, 36, 201–16.

Zihlman, A. (1978). Women and evolution, Part 2. Subsistence and social organization among early hominids. *Signs*, 4, 4–20.

Zihlman, A. (1981). Women as shapers of human adaptation. In *Woman The Gatherer*, ed. F. Dahlberg, pp. 75–120. New Haven: Yale University Press.

Zubrow, E. (1989). The demographic modelling of Neanderthal extinction. In *The Human Revolution: Behavioural and Biological Perspectives on the Origins of Modern Humans*, ed. P. Mellars & C. Stringer, pp. 212–31. Edinburgh: Edinburgh University Press.

Zuckerman, S. (1932). *The Social Life of Monkeys and Apes.* New York: Harcourt.

LIVERPOOL
JOHN MOORES UNIVERSITY
AVRIL ROBARTS LRC
TITHEBARN STREET
LIVERPOOL L2 2ER
TEL. 0151 231 4022

Index

adaptation
  behavioural, 22, 39, 71, 106
  climatic, 38–40
  dietary, 22–3
  evolutionary, 22–4, 33–4
  genetic, 27–8
  learned, 28
adaptive radiation, 34
adultery, 247, 253, 254–5
agonistic buffering, 222
Allen, N. J., 232, 285–6
alliances, 7, 10, 149, 165, 215, 242, 256, 287
  coalition, 231
  lineages and, 12, 231, 257–9, 284
allomothering, 22–3, 66, 219
anecdotes
  accumulated as evidence, 7, 179–83
  problems with as avidence, 7, 167–76
anthropological genetics, 69–71
ape language studies, 8, 181, 196
Arnhem Zoo, 149
  Dandy, 175–6
  Nikkie, 171–5
  Spin, 171–5
  Zwart, 175
artificial intelligence, 4, 83, 92, 274, 284
  expert systems, 83, 104–5, 111
  simulation models, 102–11; General
    Problem Solver, 103, 107; neural
    networks, 110; Soar, 106–10
awareness, 166
aye-aye, 34

baboon, 2, 12, 15, 16, 43, 61, 87, 146, 177, 194, 221, 222, 264, 284
  Hammadryas, 29, 46, 113
  olive, 28, 221
behavioural ecology, 2, 20, 21, 56, 57, 71, 77–8, 216
behavioural flexibility, 20, 33–4, 35, 106
behavioural variability, 4, 22
  forms of, 2, 26, 27–8
  in relation to genetic variation, 25–9

in relation to habitat, 25–6
bias
  in human evolution theory, 20–2
  about natural behaviour, 22–4
  in studying behaviour, 1–2
blocks world, 199–203
Borgerhoff-Mulder, M., 250–1, 254
brothers, 223–4
Brunner, J. S., 275
Byrne, R. W., 7, 157–61, 179, 212

Cartmill, M., 123–4
Cayo Santiago, 9, 16, 77, 168, 207, 223–4, 235
Chagnon, N. A., 237, 250–4
Cheney, D. L., 101, 142–4, 181–3, 205, 219
chimpanzee (Pan)
  paniscus, 160, 180, 264; Kanzi, 14, 277
  troglodytes, 6, 8, 10, 18, 43, 45, 61, 107, 133, 146, 151, 156, 160, 169–78, 179–181, 189, 190, 192, 194, 212, 219, 220–1, 224, 243, 264, 277, 281; Austin, 8, 195–203, 206–7; Figan, 169–71, 173–4, 175, 180, 181; Goliath, 173–4; Mike, 192; Sherman, 8, 195–203, 206–7
Chomsky, N., 13, 272, 274, 280
cognition, 5, 100, 102, 107, 117–38
  animal cognition as compared with human cognition, 189–94
  comparative approach, 264
  evolution of, 102, 110, 111, 191
  social cognition, 5, 112, 113, 116, 117, 139–64, 165–7, 175, 178, 263, 264–6, 274, 284
  unified theory of, 82, 105–7, 110, 116
communication, 179;
  and culture, 255–6
  forms of, 194–203
  functions of displays, 195
  gestural, 266–7, 270, 271
  informational theory of, 1, 4, 93, 113, 186–211
  and management of relations, 93–102

318